Your Choices Change Lives

A Marine's Journey of Rising from the Ashes

By Cpl. Todd Nicely

Your Choices Change Lives
A Marine's Journey of Rising from the Ashes

Published by Reflek Publishing

Disclaimer: While the ideas and strategies in this book are shared with the best intentions, results may vary from person to person. There are no guarantees concerning the level of success you may experience, as the goal is to offer insight, encouragement, and inspiration. To respect the privacy of those involved, some names and details may have been changed.

For more information:

Paperback ISBN: 978-1-968164-09-6
Ebook ISBN: 978-1-968164-10-2

To my family, to 2/2 Fox Company First Platoon First Squad, and to any vet out there who is struggling. Keep going.

If you are a veteran who is struggling, dial 988.

Table of Contents

PROLOGUE

ABOVE THE DIRT

The sun hadn't made it over the treeline yet, but there was enough light to see my breath hanging in the air as I adjusted my Marine Corps league uniform. These practiced movements with my prosthetic hands had become second nature over the months— straightening the collar, checking the ribbons, making sure everything was squared away. You can take the man out of the Marines, but some habits never die.

Only one car sat in the parking lot this morning. These older vets, the ones who served in Korea, Vietnam . . . their funerals get smaller every year. Fewer people remember. Fewer people show up to say goodbye. That's part of why I made it a point to be there, making sure they got the send-offs they deserved even if it was just me and a couple of other Marines from the Honor Guard on watch.

I settled into my Segway chair, feeling the weight of the ceremonial bugle in my lap. The morning air bit at my face, that kind of cold that reminds you you're alive. The silence in these moments isn't like other kinds of quiet.

Somewhere in the distance, a crow called, and just like that, I was right back there. Back in the Afghan dirt, March 26, 2010. The explosion. The feeling of being lifted into the air. That first thought when I landed: "Oh shit, they finally got me." The taste of blood and dust. The sound of my guys scrambling to save my life.

But I'm not there anymore, I thought. *I'm here. Above the dirt.*

That had become my guide over the years, a reminder and a declaration all rolled into one. Every morning when I brush my teeth, when the toothbrush hits the floor (which it usually does), when simple tasks become complicated puzzles to solve, I remind myself: *I'm above the dirt.* Some of my brothers didn't make it back. Some of them came back but couldn't find their way through the darkness. That's one reason I tell my story at FOCUS Marine Foundation . . . to help other warriors find ways to deal with their demons and know they're not alone. That's one reason I'm writing this book, too—in case I can reach more vets that way. And me? I'm still here. Still breathing. Still serving, just in a different way now.

I can't help but shake my head at the irony. All those years ago, when I was trying to get into the Marines, I had to beg them to let me in. Had to get waivers, jump through hoops, and prove myself worth taking a chance on. Now here I am, still serving the Corps, just with a few less original parts.

I had arrived early for that Honor Guard ceremony, like I always did. Being the first one there gave me a moment to settle my thoughts, push away the emotions that tried to creep in, and remind myself why I was here. I knew what it meant to the families standing on the other side of that grief. And I knew what it meant to the men being laid to rest.

Because I had come really fucking close to being one of them.

A gust of wind swept through the cemetery, cutting through my uniform. I felt it in places that no longer existed—phantom limbs, phantom pain, phantom life.

No one tells you about that part.

And then—the sharp crack of the three-volley salute.

The shots shattered the stillness, cutting through the winter air, cutting through my thoughts, through everything. Even though I had heard them a hundred times before, even though I knew they were coming, they always hit the same way.

A final send-off. A final honor.

Later, pulling away, the cemetery shrank in my rearview mirror, but the weight of it stayed with me. That's the thing no one tells you about survival—it's not just about making it out alive. It's about figuring out what the hell you're supposed to do with the days you've been given.

Fifteen years ago, I didn't know if I'd ever see another sunrise.

But to understand how I ended up in that moment, we have to go back.

Back to Jefferson County.

Back to a middle child who was about to learn the hard way that every choice has consequences.

A Message to Readers

People ask me all the time what they're supposed to learn from my story. Like there's some magic answer that makes everything make sense. The truth is, it's pretty simple: your choices change lives. Not just your own, but those of everyone around you.

It took me a long time to figure that out. I had to learn it lying in a canal missing all four limbs, sitting in a hospital bed wondering if life was worth living, and staring down the barrel of my own gun and pulling the trigger only to come out on the other side—again. Each choice led somewhere, good or bad.

The hardest lesson?

Strength isn't about how much you can handle alone. Real strength is knowing when to ask for help. That's why I'm here talking to you through this book—to prove to the next vet out there struggling that they're worth that ask.

And that there's a hell of a lot of life on the other side of it.

Chapter 1

THE THIRD CHILD

MARCH 26, 2010 - HELMAND PROVINCE, AFGHANISTAN

The morning air hit my face like a slap, that kind of bone-deep Afghan cold that somehow feels different than anywhere else. Not that I was focused on the weather. When you're checking your gear for what feels like the hundredth time, making sure your squad of Marines is ready to move out, the cold is just background noise.

I watched them gear up—most of these guys were just kids really, eighteen or nineteen years old. The sound of equipment checks and weapons being cleared echoed across our position. That familiar click-click of metal on metal becomes like a heartbeat out here. My own gear—forty pounds of body armor, ammo, and equipment—had become such a part of me I barely noticed it anymore.

"Ammo check," I called out, watching them respond in sequence. These guys were solid. I'd made damn sure of that through training. But being solid in Helmand Province isn't always enough. One wrong step, one moment of complacency, and . . .well, we all knew the stakes. I studied their faces as they worked, reading those little tells I'd learned to spot long before I ever put on a uniform. Some had that tight jaw that meant they were hyper-focused, with others trying too hard to look casual.

Standing there, watching over them, the same thought hit me that always did before we moved out: if something happens today, better me than them.

**

Life has a funny way of teaching you what you need to know, even if you don't realize it at the time. Twenty-five years before that morning in Afghanistan, I was just a kid learning to read faces in Jefferson County, Missouri—not the kind of place that made headlines.

It was small-town, a patchwork of tight-knit neighborhoods and long stretches of nothing. A place where people worked with their hands, where kids grew up fast, and where your last name could mean as much as your first.

For me, it meant learning how to hold my own.

My folks had their issues in those early days. Still, Dad taught me about hard work, I'll give him that. He had me out on construction sites when I was eight or nine, sweeping up after the drywall crews, learning how to work with my hands. My old man taught me how to measure twice and cut once, always a master carpenter. Sometimes weekends with him meant bars, but not always. Sometimes we'd go out to my grandpa's land and ride four-wheelers or play paintball, and he taught me how to shoot clay pigeons right handed. (This came in handy later.

The Marines told me I was left-eye dominant, so I had to learn how to shoot left-handed. At the end of the day, being able to shoot with both of my hands turned out to be useful . . . back when I had both hands, anyway.)

Mom—she was something else entirely. Worked multiple jobs, crazy hours. Most nights she'd come home so exhausted she'd just collapse on that old couch. Every Christmas she'd go into debt making sure we had presents under our tree. Looking back now, I understand she was trying to give us everything she never had, even if it meant working herself to the bone.

They got divorced when I was in third grade. It shouldn't have been a surprise, but it still hit like a truck. We moved into a cramped apartment then, and our living situation wasn't exactly what you'd call ideal. Mom slept on the couch—one of those old ones that had given up trying to stay firm years ago—in the living room. Four of us boys crammed into one bedroom, a jungle of bunk beds and barely-contained chaos. My two sisters had the other room, though having your own room didn't mean much when seven people shared one bathroom.

Being the middle child of six kids, you learn things. You have to. With older siblings blazing trails of trouble ahead of me and younger ones needing looking after behind, I developed a sort of sixth sense about people. I watched how my brothers and sisters interacted, and learned to spot when a fight was brewing just from how someone walked into a room. These skills came in handy later in Afghanistan . . . turns out, there's a hell of a lot more to communication than just words that come out of people's mouths. I was excellent at reading body movements, shoulder shrugs, and all-around physical demeanor. Sometimes, I would go to the same house my interpreters had just left to see if I would get the same reaction out of the occupants, ensuring nobody was lying to me. I could always tell. When I was young, it usually fell to me to keep the peace between the younger kids. Not because anyone asked me to—it just seemed like that's where I fit. When I was wearing a uniform, though, the stakes got a whole lot higher.

Later, when my stepdad came into the picture, we moved from that cramped apartment to a trailer with him. It wasn't just a change of address—it was like someone finally opened a window and let some fresh air in. He turned out to be more of a real father figure than I could've imagined back then.

You know what's crazy? All of that—the crowded apartment and trying to keep the peace between siblings—was actually preparing me for that morning in Afghanistan. Learning to read people, understanding what responsibility means, and seeing how one person's choices ripple out to affect everyone around them.

Looking back, I was probably what you'd call a handful. A natural athlete, sure, but with this streak in me that always wanted to push limits. When I was four or five, our neighbors (who were gymnasts) taught me how to do a backflip on our trampoline. That was it. I was hooked. Pretty soon I was doing backflips off of anything I could find: trailers, decks, cliffs. Understandably, it drove my mom absolutely crazy. She'd get so scared watching me, but I couldn't help myself. If it had height to it, I wanted to jump off it. When I was 20 years old, I did a backflip off a friend's trailer, and my foot landed in a hole, snapping all five of my toes. I had a cast on for about three months. The day I got it off, I went to a friend's wedding rehearsal dinner, and we went back to their house for drinks afterward. I set up a ramp on a railroad tie with a mound of dirt that made the whole thing about seven feet high. You might guess what happened next: I tried to do a backflip on a bike, came down, and snapped my ankle. I wasn't even without a cast for a full day.

I guess you could say anything I was told not to do back then, I did anyway.

I started playing football at seven, and that became my thing.

Something about the game clicked for me. Maybe it was the combination of physical challenge and having to think three steps ahead. Or maybe it was that I finally had somewhere to put all that energy always churning inside me.

School was . . . complicated. The teachers would say I had "academic potential," whatever that means. They kept testing me for advanced classes, and I'd do well enough to get in, but I never quite made it stick. It's not that I couldn't do the work—my brain just always seemed to be somewhere else. Like I was restless for something, though I couldn't have told you what.

Quick wit got me into trouble as often as it got me out of it. I developed a knack for solving problems, but not always in ways the adults around me appreciated. Risk-taking? Yeah, you could say that was my middle name. Though maybe "reckless" would've been more accurate back then.

Football was where I first started to understand what leadership meant. It wasn't about being the loudest or even the best player—it was about reading the field, understanding what needed to happen next, and somehow getting everyone to move in the same direction. Really, I was just doing what came instinctively.

But home was where I learned about responsibility. Being a middle kid meant I was old enough to watch the younger ones but not old enough to do my own thing like the older kids. By nine, ten, and eleven years old, I was keeping an eye on my younger siblings. We had cleaning rotations—whose turn it was to do what chores—and somehow I usually ended up being the one making sure it all got done.

The thing about family tensions is they teach you how to navigate difficult situations. One minute you're trying to keep your younger siblings from killing each other over who gets the last Pop-Tart, the next you're mediating some complex dispute about who borrowed whose clothes without asking. You learn to read the room, sense when something's about to blow up, and sometimes, if you're lucky, how to defuse it before it does.

My relationship with my dad was complicated. Construction work taught me the value of hard labor—I mean, there's nothing quite like being on a job site and hanging drywall when you're in middle school. Hard work? Yeah, he definitely taught me that. But weekends with Dad were a different story. You don't really understand as a kid that what's happening isn't normal. You just know to watch for the signs—how someone's standing, the tone of their voice, or the way tension can fill a room faster than cigarette smoke.

Those lessons about reading people came with a price.

Life with Mom and my new stepdad Mikey felt different. Mikey taught me more about life than he ever had to—how to be grateful for what you have, how to carry yourself, how to be there when people need you. He took on six kids who weren't his responsibility, showing us what it truly means to take care of people and make sure they have everything they need . . . even when *you* don't. When we were young and dumb, he would let us borrow his trucks, and we would wreck them. Still, he'd fix them and let us borrow them again. Mikey was a man of integrity who kept his word. Even when he was mad at you, he wasn't afraid to help you fix your brakes or come pick you up when your car broke down. Even though we weren't his kids, we *were* his kids. For those reasons, the move from the apartment to the trailer felt like entering a different world. Sure, Mikey worked nights, but anything you asked of the man, he'd do it for you. That's the crazy part: as much as I rebelled against both him and my mom (and believe me, I did), they were always there for me.

But families are messy, and love often comes wrapped in conflict. My poor mother. As much as we've butted heads over the years, I can never forget the love she had for us.

**

You hear people talk about how your whole life flashes before your eyes in moments of crisis. That's not exactly true. What really happens is you start seeing connections—how one thing led to another, how every choice pointed you toward this moment, how those choices change lives—mine, Mikey's, Dad's, Mom's . . . the list goes on.

Being a squad leader in a combat zone isn't something any training manual can really prepare you for. Yeah, they teach you the tactics, the procedures, and how to call for fire support or medevac. But nobody can teach you how to carry the weight of knowing eleven other lives depend on your decisions. That's something you have to figure out on your own. That's something you have to learn to live with, and no one really talks about how it destroys your psyche. It also destroys your hope. It destroys everything about you . . . so you have to figure out how to be numb to the feeling to get the job at hand done. You have to learn to be numb to the simple fact that if something goes wrong with your mission, it all falls on you.

I had "Your Choices Change Lives" tattooed on my body in between deployments, and I was about to understand that phrase in ways I never could have imagined. Sometimes life doesn't just test you—it detonates under your feet and blows apart everything you thought you knew about yourself.

But first, I had to learn *some* choices. Turns out those can land you in front of a judge.

Chapter 2

FINDING DIRECTION

MARCH 26, 2010 - HELMAND PROVINCE

The maps of Helmand Province were spread out before me. Grid after grid that had become as familiar as the back of my hand. Behind me, my squad was going through their final preparations. The morning's mission would be straightforward enough—a standard patrol to check markers. The lieutenant wanted to come along and talk to some locals.

Just another day in the sandbox.

I traced our planned route with my finger, muscle memory from countless briefings kicking in.

But my mind kept returning to a moment, years earlier, when the idea of being here and responsible for these lives would have seemed about as likely as me becoming an astronaut.

**

At sixteen, I was well on my way to becoming either a statistic or a cautionary tale. Maybe both. The school district had redrawn its boundaries, which meant all my friends went to one high school while I got shipped off to another. New place, no crew, no anchor except for football—and I'd already managed to screw that up by getting expelled in eighth grade for drinking on the school bus.

It wasn't my brightest moment. Some girl and I decided getting drunk on the way to school seemed like a good idea. It wasn't. Missing freshman football hit me harder than I expected. Football had been my constant since age seven, the one place where things made sense. Without it, that first year of high school felt impossible.

By fourteen, I was working as a fry cook and pushing carts at the local grocery store by sixteen. The paychecks gave me a taste of independence that probably went to my head. I bought my own car, which led to epic showdowns with my mom. At the very wise age of 16, I thought that a car would always give me independence . . . being able to take girls on dates, being able to open up in high school . . . you get the picture. Believe it or not, I was very shy and had low self-confidence. I always believed that everybody thought they were better than me. However, now I know that was just my projection of their thoughts because, by my senior year, I was always at parties—usually doing backflips to show off. Being in a new high school with no friends to start with took a toll on me, and I always thought I was "behind." Looking back, I know I had friends . . . but I just didn't always see it. And that car? That was my ticket, or so I thought.

It also caused a hell of a lot of problems.

"You can't take my car," I told Mom once when she tried grounding me. "I bought it myself."

"I can, and I will," she shot back, holding my keys hostage.

"I'll call the cops and tell them you stole it if you don't give me my keys back."

I was a shithead before I joined the Marines—those are my words, not anyone else's. I absolutely thought I was invincible just because I worked for what I owned. Never mind that she still put food on the table, kept a roof over my head, and did all the thousands of things a mother does. I couldn't see past my own pride.

Pride. It's funny how that works. The same thing that made me stand up to my mom back then would later help me push through Marine training and keep me going when everything else was stripped away. The same quality that helped me thrive in the Marines, the "we take care of ourselves and our own" mentality, is also what hurt me because I couldn't ask for help when I really needed it. But I didn't know any of this yet. Back then, pride was just making me impossible to live with. The more my family tried to rein me in, the harder I pushed back. Classic teenage rebellion, maybe, but I was taking it to new levels.

Then came the night that changed everything. At sixteen years old—me, my three brothers, and one of my friends—were watching *Saving Private Ryan* at the theater. Every scene, every moment had me mesmerized by the fact that these men were willing to give up their lives for each other and be so selfless. The whole movie changed my life, and the D-Day scene in particular hit me like a physical force. Those men charging into certain death, willing to give everything for something bigger than themselves. I had to pee so bad my legs were shaking, but I couldn't move. Couldn't get up. Couldn't look away.

Something clicked at that moment. If these guys had the audacity, the pure balls to put their lives on the line for a country they believed in, I owed them something. I owed them my best shot at keeping what they'd died for alive. For the first time in my life, I had an idea about my future.

It wasn't just the violence or the action—God knows I'd seen plenty of war movies before. It was watching men choose to do something impossible because it needed doing. Watching them push through their fear because others were counting on them. Something in that spoke to a part of me I didn't even know existed.

By senior year, I had it all figured out. I'd take my ASVAB, join the Army, and find my purpose—so I started the whole thing in motion. The recruiter was all set to come by, and all I needed was my mom's signature because I was only 17. Then 9/11 happened—I still remember it clearly, the teacher wheeling the TV into our drafting class, us sitting there glued to it, watching the second tower come down, with our mouths hanging open in shock. After that, three days before my recruiter was scheduled to show up, my mom put her foot down: she wasn't signing. No amount of arguing or pleading would change her mind. Looking back now, I get it. What mother wants to sign papers that could send her son off to war? But at the time, it felt like having a door slammed in my face. Bad things seemed to happen to me in three-day windows; three days before I was supposed to get my braces off in high school, I ran into the bleachers and knocked some of my teeth out. Three days before I was set to re-enlist in Afghanistan, I stepped on the IED. I guess I've never had a sense of self-preservation . That lack would serve me well in combat, allowing me to run toward danger without a second thought.

When my mom refused to sign and derailed my plans, I was pissed because I knew it was something I was meant to do. But I was also a young man, and I'd met a girl who said she wouldn't stay with me if I left. That dulled the sting for a while, until that relationship fizzled and I was right back in my own head. Looking back, I'm not upset that my plans had to change because who knows where I would've been. Maybe I could've been a casualty, or maybe I could've been a Gunnery Sergeant or a Staff Sergeant by then. All I know is God put me where I needed to be, and that was bringing home all of my men on that day of March 26, 2010.

After I graduated, I did what seemed like the next best thing—went to work construction with my old man for almost five years. The money was good, damn good for a kid my age. But something was missing. That gnawing emptiness I'd felt watching *Saving Private Ryan* never went away. Meanwhile, Jefferson County was being ravaged by heroin and methamphetamine. Friends were dying from overdoses. Many of the ones still alive were shells of themselves, headed nowhere but to the next fix. Temptations and opportunities to make poor choices swarmed us like flies on shit.

I watched guys I'd grown up with disappear into that darkness. It was like a slow-motion train wreck you couldn't look away from. Every story about another overdose, every friend lost—they all whispered the same thing: *Get out. Get out before you end up addicted to heroin or in prison.*

But getting out isn't as simple as just deciding to leave. Life has a way of trapping you in patterns, in habits, in the comfortable misery of the familiar. I was making good money in construction, renting a place with a couple of my buddies, and could buy my own beer. From the outside, I probably looked like I was doing alright. Inside, though, the emptiness just kept growing. And it only got worse when the house burned down and I had to move back in with my mom. I'd gotten used to independence, and here she was, telling me what to do again. We clashed more times than I could even count, resentment brewing. By 2006, I was spending most of my money at bars, watching my life circle the drain one drink at a time. Then came the night that changed everything. I was sitting on a bar stool with my buddy, both of us half-drunk and going nowhere fast. We looked at each other, and I asked the question that had been eating at me:

"What the fuck are we going to do with our lives?"

He just stared into his beer. Neither of us had an answer. The construction work with my dad had dried up—2008's economic crash was already starting to bite. Friends were dying, falling into addiction, or just . . . fading away. I could feel myself sliding toward something dark, and for the first time, that scared me more than the idea of change.

"I'm thinking about joining the Marines," I said.

The Marines. Not the Army this time. If I was going to do this, I was going to go all in. Pick the hardest path. Challenge myself in ways that scared me. My buddy said he'd do it too. He ended up going in before me and landed at Okinawa, never making it to war. Today, he lives with survivor's guilt because I know he didn't want to leave me behind. But it wasn't his fault. Turns out I had some things to take care of first.

The "things to take care of" were DUIs. Multiple ones. The judge looked at my record and saw two paths: five years in state prison, or sixty days in county with a chance to turn my life around. The deal was simple—serve the sixty days, then follow through with my Marine Corps commitment.

Some might call it luck. I call it the wake-up call I needed. Sitting in that county jail cell, I had plenty of time to think about choices. About how each decision ripples out, affecting not just you but everyone around you. I thought about those men storming the beach in *Saving Private Ryan*, about the friends I'd lost to drugs, about my mom refusing to sign those Army papers years ago. I dreamed about the Marine Corps, which pushed me through. I even started my own little training camp by playing cards and betting people for pushups. The guys in the clink were all rooting me on, always saying I had the potential to be a great leader. That might be hard to swallow coming from a bunch of convicts, but if someone can see something positive in you that you can't see in yourself, your best bet is to listen.

Sixty days gives you a lot of time to get your head straight—or at least, you'd think. When I got out, I was looking forward to boot camp but fell back into my old ways. Construction had been dried up for a while thanks to the economic crash, so I went back to working for my buddy's dad at a steam turbine shop. It was beyond clear that I needed to get out of there before I landed myself back behind bars. I got the call two months after I got out of jail that they weren't going to give me an opportunity to mess up: I was going. They were shipping me out, bypassing the poolee system.

I needed to get on that plane to boot camp as soon as possible. Mom was still worried, Dad was still Dad, Mikey was still proud, and I was ready to get the hell out of there.

**

That day in March, standing over those maps in Helmand Province, I watched my men finish their weapons checks, some listening to pump-up music through their earbuds. They were tired—we all were. This deployment had been grinding us down, day after day of patrols, firefights, and that constant tension of knowing every step could be your last. But they were solid. They were Marines. They were *my* Marines.

I saw reflections of my younger self in some of my squad. The same cockiness, the same need to prove yourself, the same search for meaning—working with them was a taste of my own medicine. But I also saw something else, something I'd learned back in Jefferson County: your choice changes lives. Not just your own, but everyone around you.

I took that first step forward, not knowing it would be one of my last on my own two feet. But that's the thing about choices—you never really know which ones are going to change your life forever until after you've made them. I adjusted my weapon and moved out. Whatever was coming, we'd face it together. That's what Marines do.

To learn that, we have to go back to when I discovered what it truly means to lead from the front.

Chapter 3

LEADING FROM THE FRONT

MARCH 26, 2010 - HELMAND PROVINCE

The bridge looked like shit. Not much more than some rotting planks laid across a narrow canal, but we had to cross it. Wind kicked up dust around our boots as my squad got into position. I scanned the far bank through my scope—nothing but mud walls and empty doorways. Perfect place for an ambush.

Seven Marines waited behind me. Good kids, most of them 18 or 19. I knew each one by now: Thomas, who was carrying the radio that day instead of me, thank God, and probably plotting his strategy to beat me the next time we played chess. Camarillo, who was always on time and had his gear in order. Harrigan, who I called "my shaker" because he sometimes trembled. The weight of their lives sat heavier than the 40 pounds of gear on my back.

The lieutenant wanted us to check some marker and talk to a local shopkeeper. A simple mission on paper. But the shop that usually buzzed with activity sat empty. Even the local kids who'd beg for candy were nowhere to be seen. In Afghanistan, normal meant safe. This wasn't normal. Even the usual morning bullshitting had died down to whispers.

The air had that stillness to it, like the moment before a storm hits.

**

By the time I got to boot camp in San Diego, I knew two things for certain: I was done with my old life, and I was going to be a Marine. The Corps had signed me up despite my record. I had to get waivers and jump through extra hoops. While other guys were getting bonuses to enlist, I was basically begging them to let me in.

I remember standing on those yellow footprints at MCRD, getting screamed at by drill instructors, wondering, *What the hell did I just get myself into?* There I was, a guy who didn't like to be told what to do, getting told what to do every fucking second. We got yelled at constantly, but that was part of the deal. We could never do anything right. My gunny at FOCUS Marine Foundation (where I mentor today) says it best: they train commitment despite consistent failure. Looking back, I see that's what it was—teaching you to never turn on your fellow Marines even if you're the one getting in trouble because the guy on your left or right did something stupid. The physical side wasn't too bad because I was a natural athlete and in shape, but it took me a while to get used to the rest of it. As camp went on, I stepped into a leadership role. If I saw a fat guy getting called out and made to do extra pushups, I'd go up and do them with him so he wasn't alone. I decided I wanted to become a sergeant in the first four years—something unheard of, but I thought, *What the hell? I'm here, and I'm all in.*

As boot camp went on, something was changing in me. I could feel it. That wild kid who used to do backflips off cliffs was learning about discipline. That shithead who threatened to call the cops on his own mother was starting to understand respect. That lost soul on a bar stool was finding purpose.

The hardest part wasn't the physical part, though God knows there was plenty of that. I remember getting up in the middle of the night a lot to jump rope, figuring it would improve my run time. The senior drill instructor thought I was crazy, but I did it anyway. I could handle all that. But you don't just wake up one morning completely changed. It's a thousand little decisions, day after day. Running when you want to sleep. Eating right when you want to grab fast food. Staying home to train when your friends are out having fun. Each choice either reinforces your new direction or pulls you back toward the old one.

But something weird happens when you start making better choices—they get easier. That discipline you're forcing on yourself starts to feel natural. The more I prepared, the more I could feel myself becoming someone different. Not just physically stronger, but mentally tougher. That kid who used to rebel against any kind of authority was learning to channel that energy into something productive.

Next, I went to Camp Pendleton in California for the School of Infantry. During conditioning hikes, we'd haul every weapon system known to man up a monster hill. M249s, M240s, mortars—if it could kill, we carried it. One day, a fellow Marine was struggling with a SAW and full combat load. I grabbed his weapon and ended up humping two machine guns while he held onto my shirt for the last mile. There, one of the instructors pulled me aside one day after a particularly brutal hike and said: "You keep looking out for others, Nicely. That's what leaders do." Coming from him, that meant something. He was old school Marine Corps, and didn't hand out praise easily. He showed me there was a different way to lead . . . not through fear, but through example. The training got more intense as we prepared for deployment. The days were long: live-fire exercises, combat scenarios, and endless drills.

The final stop before deployment was Fleet Time at Camp Lejeune. By that time, I was 23 . . . an old man by Marine standards. That training kicked my ass, but not in the ways I expected. Again, just like in boot camp, the physical stuff was fine. It was watching poor leadership there that really got to me. Our squad leaders were mostly dick-swinging teenagers drunk on their first taste of power. They'd bark orders without explanation. "Shut up and do what you're told." No context, no understanding of the mission. Just blind obedience. I'd catch myself thinking, *Why the hell are we doing this?* But in the end, it was worth it. My favorite part of Fleet Time was our time in the field bonding with the group—figuring out who was who, what our roles were, building trust, and connecting before the storm.

One of those late nights in North Carolina, I went out with a couple of buddies from the platoon. We had a few drinks, and they all headed home. Me? I headed to the tattoo shop, where I got "Your Choices Change Lives" inked on my back, complementing the two theater masks already there—one symbolizing the dark, the other symbolizing the light. Even though I'd had a little to drink, I didn't regret it at all: choices *do* change lives, especially when we're talking about war. When my buddies woke me up the next morning, they didn't know what the hell happened. With the bandages on my back, they thought I somehow went out and got myself burned the night before.

Turns out, that tattoo would mean more than I ever thought at the time.

**

I'm not going to lie—I thought my first deployment would go way differently than it did. In Iraq in 2008, our unit was part of the team building schools and infrastructure. It was seven months of watching a bridge near Rawah, with occasional mortar rounds landing nowhere important. We spent our days doing security patrols, watching that damn bridge to make sure gas and supplies could get where they needed to go. The whole thing felt like a practice run compared to what was coming. The most action we saw was breaking up fights between local kids trying to steal our water bottles. But that deployment taught me more about leadership than any training course could—and I learned a lot of it by watching Corporal Roark.

I watched how he handled his firearms, how often he raised his voice, and why. I watched how he was the first one up, the last one to eat, the first one to volunteer for shit detail. I paid attention to how he explained the why behind every mission. Leaders like that? Guys would follow them anywhere. We knew Roark cared about keeping us alive more than looking good for brass.

Then there were the other types. Leaders who'd hide in the FOB while sending their men out on patrol. Guys who treated their Marines like disposable parts instead of brothers. You could spot their squads a mile away—no cohesion, no trust, just a bunch of scared kids following orders. I saw how that kind of leadership could get people killed.

The deployment might have been quiet, but it showed me exactly what kind of leader I wanted to be and what kind I didn't. I spent those months studying everything: how to read terrain, how to position men, and how to build trust. When things occasionally did pop off, I noticed how the good squads moved like machines while the bad ones fell apart.

When we got orders for Afghanistan in late 2009, I had my own squad. Twelve Marines looking to me for direction. The responsibility terrified me, but I couldn't let it show. Leaders don't have that luxury. Everything I'd learned was about to be tested in ways I never imagined.

From day one with my squad, I threw out the standard playbook. No more blind orders or "because I said so" bullshit. Every Marine needed to understand not just what we were doing, but why we were doing it. The other squad leaders thought I'd lost my mind.

"You're giving away too much intel," they'd say. "Just tell them where to point their weapons." But I remembered that scared kid at Lejeune, following orders that made no sense. My guys deserved better. If they were going to trust me with their lives, they needed to know the whole picture.

I ran training differently than anyone else in the company. I had my guys practicing skills way above their pay grades—calling in nine-lines for casualties, coordinating air support, and planning tactical responses. The brass hated it. "That's officer shit," they'd tell me. But I knew if I went down, my guys needed to know how to carry on.

Even our patrol formations broke from tradition. That Ranger file bullshit they taught us— walking in straight lines like ducks in a shooting gallery—wasn't happening with my squad. I drew from playing paintball as a kid. I kept us spread out, using terrain for cover, and always having overlapping fields of fire. The other squad leaders called it cowboy tactics. I called it staying alive. I figured we weren't fighting a conventional war, so why should we use conventional tactics?

Every night after training, I'd drill my guys on scenarios. What if we lose comms here? What if we take fire from this direction? What if our primary exit gets blocked? They'd groan about it, but they did it. Pretty soon they started coming up with their own scenarios, thinking three steps ahead. That's when I knew we were getting somewhere. The whole point was to come back alive, and I was going to do whatever I could in my power to make sure that happened.

Getting married during my service changed things. I'd met my first wife Crystal before Iraq—we started as pen pals while I was deployed. Writing letters back and forth, you really get to know someone. There's no awkward face-to-face stuff, just putting your thoughts on paper. We got married in February 2009, right before Afghanistan.

The responsibility of being a squad leader hit differently after marriage. These weren't just my Marines anymore—they were somebody's sons, brothers, husbands. Every decision I made affected not just them but their families back home too. That weight never went away.

Building trust with my guys didn't happen overnight. From day one, I threw out the standard playbook. My approach was simple: never ask someone to do something you wouldn't do yourself. Sounds basic, but you'd be surprised how many leaders forget that. Whether it was humping extra gear on a ruck march or running point through Taliban territory, I made sure my guys saw me out front.

Leading from the front wasn't just about being first into danger. It meant making sure my guys had what they needed before I took care of myself. If someone was short on socks, they got mine. If we were low on water, I would go thirsty first. You're only as good as your team. When leading from the front, I knew every man behind me deserved everything *before me*. Little things like that build trust.

When they made me squad leader, training ramped up hard. We had to be ready for Afghanistan, and everyone knew it would be different from Iraq. The more we trained together, the more I saw these guys becoming a real unit. They started looking out for each other without being told, thinking three steps ahead, working as one.

By the time we deployed, I had twelve Marines counting on me to bring them home.

But all that training and preparation—like making sure your gear was clean, your weapon was clean, or what to do in different scenarios—sometimes doesn't mean shit. Standing there on that bridge in Helmand, something felt wrong. The small outside shop that sort of looked like a storage unit and sold snacks was empty, and it usually had at least a few local kids begging for candy. Something was off.

The Corps teaches that how you lead determines who follows. Through all the training, the deployments, and the firefights—that lesson stuck with me. Being a Marine wasn't just about fighting. It was about taking care of your people, earning their trust, and bringing them home.

Every choice ripples out, affecting not just you but everyone around you. Those early lessons of watching good and bad leaders shaped how I'd handle my own squad. The examples set by guys like Corporal Roark showed me what real leadership looks like. That morning on the bridge, the fundamentals stayed the same: lead from the front, take care of your people, and complete the mission. Everything else is just details.

My guys trusted me with their lives. That trust was earned through countless hours of training, through showing them I'd never ask them to do something I wouldn't do myself. They knew if things went sideways, I'd be right there with them. That's what leadership means: being willing to take the same risks you ask others to take.

Sometimes I think about that morning, about all the little choices that led us there. One foot in front of the other. One step at a time. Sometimes that's all you can do.

Chapter 4

BROTHERS

MARCH 26, 2010 - HELMAND PROVINCE

Earlier that month on my birthday, my squad and I were sitting around playing Spades and passing warm Gatorade back and forth, talking about what we'd do when we got home.

(You can probably imagine what came up in that conversation.)

It wasn't much of a party, but out there, any excuse to bullshit and forget about the war for a minute was worth it. None of us knew how much everything was about to change.

I was 26 then, ancient by Marine standards. After over 200 firefights in those months, we'd gotten tight as a unit. I knew what each of them was probably thinking that morning as we started across the bridge.

My fingers found their way to my wife's picture and the Hot Wheels cars my cousin gave me before I left, my lucky charms tucked in my right shoulder pocket. I'll never forget what that little guy said when he handed me those tiny cars: "These will keep you safe."

That was about to be a pretty fucking tall order.

**

When we first landed in Helmand, I knew it was going to be different from Iraq. Command stuck us way down south where no American forces had been before. The Taliban had every inch of ground dialed in—each canal, every mud wall, all the places to stage an ambush. Within days we were taking contact, sometimes twice a day.

There was tons of anxiety, especially because we knew what we were getting ourselves into. The battalion before us had become "combat ineffective," meaning they didn't have enough men—due to either being wounded or killed in action—and were too thin. So, they sent us in. The pressure, the anxiety, the unknown . . . the sheer weight of knowing the consequential nature of every fucking choice I was about to make . . . was overwhelming.

That first real wake-up call came when a local brought his kid to our OP. The boy had been hit in the head during a firefight. Part of his brain was coming out. I remember holding that wound while our corpsmen worked, talking to the kid even though he couldn't understand me. Right then I knew: you either learn to compartmentalize that stuff or you break. Some guys couldn't handle it. The ones who could got stronger.

**

November 22 changed everything. We were about ten days in, and that morning, Lance Corporal Nick Hand and I were building a toilet area for the guys, just normal stuff. He was ferocious, even at 120 pounds soaking wet. We used to mess with him, but he always got his revenge. He reminded me of a lion, and I always looked up to him even though we gave each other a healthy amount of shit. I have a vivid memory of Nick, one of many: one night, right after we landed in Afghanistan at Camp Leatherneck to acclimatize before they sent us south, we were walking home from the chow hall.

"You know this is going to be a rough deployment, right?" Nick said. "I hope I don't get killed at the end after making it all the way through."

"Yeah man, I get that," I said. "If I get blown up, I hope I don't lose my hands. I'd rather just lose my legs."

It turns out, unfortunately, we were both be predicting our own fates.

That day in November, Nick was heading out with One Charlie and said, "I'll be back, I think . . . " Those were the last words we ever spoke.

I was on radio watch when "Fallen Angel" came over the comms. Those words hit you in the gut. Then they gave his initials and everything just . . . stopped. One Bravo was trying to get to One Charlie, One Charlie was trying to get back to One Bravo. It all went sideways so fast. Nick was the point man—first into any situation. That's just who he was. Always volunteering for the hard jobs.

He took a round to the head during an ambush. Clean shot.

I knew the medevac was coming. Prior to the helicopter landing, my guys all made an agreement about which four of us were going to take him on the plane—three of us who were his best friends, and Sergeant Lattimier. But when the time came, one of my best friends was pushed out of the way by a staff sergeant—who had no ties or anything to Nick—and said, "I'm taking the spot." That pissed me off to no end, but there was nothing I could do.

The medevac pilots stood at attention while we loaded him, even while taking fire. Wouldn't move, wouldn't flinch. Just stood there saluting another fallen brother while bullets cracked past. That image stuck with me. That kind of respect goes beyond branch of service, beyond rank. Real brotherhood isn't about the uniform. It's about honoring what we're all there for. From then on, we called that place OP Hand.

I called my mom that night. She asked how things were going. I lied and told her everything was fine. How do you tell your mother that, ten days in, we'd already lost someone? That we had six more months of this ahead? Some things you just keep to yourself.

That night we played Nick's favorite songs through our iPods and told stories about him. The guys needed that closure. But sunrise came, and we were back on patrol. That's war; the mission doesn't stop for grief. You keep moving or you die. There is no middle ground.

That was the last time I would ever see my friend, and knowing that I couldn't be at his funeral killed me the whole deployment. To this day, every twenty-second of November, we go as a platoon to visit his mother to comfort her and make sure she gets new stories and memories of her son. Then, we go to his gravestone.

Nick is proof that one person can change the lives of many. His life meant so much, and even in his death, he keeps bringing people together. When I had my son years later, I made sure his middle name was Nicholas.

His death was the hardest for me to deal with in Afghanistan because I knew I had to push on. I knew I had to let my guys know *Hey, we are still here, and we still have a mission to continue.* It needed to happen in the moment, but I realized after the fact that it didn't allow me to focus on or feel my own grief and anger at losing him.

After losing Nick, leadership took on a whole new meaning. Every patrol felt like Russian roulette. You never knew if today was the day somebody wouldn't come back. The Taliban started shifting tactics too. Less direct firefights, more IEDs. It made my guys jumpy. Hard to blame them when every pile of trash could be hiding something meant to kill you.

I had to make calls that probably looked crazy from the outside. Once, we were pinned down by a machine gun team and couldn't even get our heads up without taking rounds. My guys were stuck . . . we couldn't move forward or back. I looked at them and knew if I didn't do something, we were all dead. I had to let the rest of the squad know where the enemy was, which meant I had to make a big move. *I've lived a longer life,* I figured, *so if anyone is going to go down, it might as well be me.*

I knew what I had to do, but that doesn't mean I was excited about it. I took off running across open ground to draw fire, rounds spitting at my feet. The whole time thinking, *I don't want to fucking do this.* But that's what leadership means sometimes. Doing the thing nobody wants to do because it needs doing.

We got credit for six Taliban that day. Never saw the bodies. They were good at dragging their dead away quickly. But we all got out alive. Sometimes that's the best you can hope for.

The guys started to change after Nick died, too. They got harder, and more focused. The playful bullshitting died down. But they also got tighter. You'd catch them checking each other's gear without being told, making sure nobody missed anything.

They'd play stupid games on the radio during patrol, like "Would You Rather" or the alphabet game. It used to drive me crazy, but looking back, I get it. You find ways to stay human.

Me and my buddy George Geddis, a machine gunner on my team, had our thing after firefights—we'd listen to Colbie Caillat's "Bubbly." Sounds weird as hell, but it was our feel-good song to bring the adrenaline down. You find what works. These guys became more than just Marines in my squad. They were my brothers. When you're stuck in trenches together, taking fire, sharing MREs, and dealing with all the shit that comes with war . . . you build bonds civilians can't understand.

I trusted my guys with my life, and they trusted me with theirs. That kind of trust doesn't come from rank. It comes from being there, day after day, going through the same shit together. Like my shaker Harrigan who got nervous during firefights. Some leaders would've written him off and kept him back. Instead, I worked with him, found ways to use his strengths. He ended up being one of my best Marines. Later, he was one of the ones who saved my life—the first one to get to me and start applying tourniquets when I was ripped to smithereens and bleeding out in the sand.

**

Being a leader means you can't show when things get to you. I had to be the steady one, the rock they could count on. After firefights, I'd find myself lying awake, running through every decision. Did I miss something? Could I have done something differently? The responsibility never lets up.

The brass wanted reports after every contact. How many rounds were fired, the enemy positions, battle damage assessment, all that stuff. But how do you put into words what it's like to lose someone? To watch a kid who was joking with you that morning get zipped into a body bag that afternoon? That stuff doesn't go in reports.

I had a saying: "Better me than them." I thought about that a lot, especially after Nick. As a leader, you're responsible for every one of those lives. Their families trust you to bring them home. Their mothers trust you with their sons. That weight sits heavy. You learn to carry it, but it never gets lighter.

By that morning in March, command had started requesting my squad specifically for missions. We'd gotten a reputation. The other platoons would be stuck in a firefight, and they'd ask for us by name: "Send out Nicely's squad." It got to the point where we were skipping over QRF and being called straight to trouble spots.

My guys were worn down. We'd been going harder than any other squad, getting called out even on our rest days. That morning I'd actually told the lieutenant we needed a break. But One Bravo needed eyes on a marker they were tracking, and somehow it became our mission.

It was weird being the guy everyone counted on. Back in training, I never would've imagined other platoons specifically asking for my squad. But responsibility has a way of finding you. After a while, the brass even started sending us out to test how other squads would handle contact, like they were using us to judge if the newer guys were ready.

I made it a point to know about everyone's situation: who had a new baby, who had a sick parent, whose girlfriend was giving them trouble. Not because I was trying to be their friend, but because that stuff affects how a Marine performs in the field.

Once, we finally got a run of packages from home and decided to make goulash, but not the way you're thinking: we made a homemade fire pit in an old bathtub and dumped the random cans in a pot on top of the fire. We played games of Risk in between patrols—which is a little ironic, I admit. We also shared everything—from socks to stories.

Another time, my friend got a beer in one of his packages.

"Corporal, can I drink this?"

"Hell man, do what you've gotta do," I said. "I didn't see anything."

I was one of those leaders who always made sure my guys had the time to do what they wanted but also also made sure they were prepared for what came next. One time, we were the furthest south we'd ever been and hadn't seen high command in a while, so I figured I'd let them let blow off some steam. One of them didn't wanna wear pants that day. George's crazy ass grew Wolverine chops on his face. You get the picture. Then, Gunny came down and saw it all.

"Nicely, what the hell is going on here?"

"These guys could all die tomorrow. Who gives a shit?"

"Get it all cleaned up because higher command is headed this way."

We did, of course—but for that little bit of time, they were their old selves fucking off, laughing together. The only thing is, we were in the desert in a war, not in somebody's garage or backyard. That's the kind of stuff that bonds you . . . not just the firefights, but the human moments in between.

Another time, it was my anniversary with Crystal. I wanted to call her, but that would mean moving our position, and I sure as hell didn't want to potentially lead my guys into combat just so I could talk on the phone. We had a big conversation about it over the radio, and gave me so much hell that they basically forced me—and *all of us,* by default—into it. After we made the trip and got to the phone, I had a total of ten minutes. I used two minutes to tell her I was alive and wish her a happy anniversary, and then I gave the other eight minutes to my guys. After all, they'd risked it just to get me there—and come to find out, they'd risk a hell of a lot more to save me later on, though I didn't know it then.

The truth about war is that it's unlike any kind of bonding experience you could ever go through in civilian life. I got to know those guys better than my own family, and in the front of my mind was always one thing: *I've got to get them what they need, and I've got to get them home. That's all that matters.*

Looking back now, I can see how everything was building to that morning on the bridge. All the training, all the firefights, all the bonds we'd built . . . it was like we were being prepared for something bigger than we knew. The guys had this quiet confidence about them. Not cockiness, but the kind of steady calm that comes from knowing exactly what you're capable of.

I remember watching them gear up that morning, thinking about how far we'd come. Some of these kids who could barely handle their weapons in training were now moving like seasoned Marines. They'd learned to read each other, to anticipate what needed to be done without being told. More than that, they'd learned to trust not just me or each other, but themselves.

That trust was about to be tested. In the Marines, you never know which day is going to change everything. You just wake up, check your gear, and the rest is up to fate, God, or whatever you want to call it.

That Choices Change Lives tattoo? I never knew just how true those words would turn out to be.

Chapter 5

THE LAST PATROL

MARCH 26, 2010 - HELMAND PROVINCE

"Oh shit, they finally got me."

The blast caught my left side and sent me airborne, which probably saved my life— though "saved" means something different when you're talking about losing all your limbs.

I landed near the canal, water splashing up on my face. If I'd gone in that rushing water instead of beside it, that would've been it. The irony is that I'd taken apart IEDs before—found them myself and dug them up. I was actually surprised one hadn't gotten me sooner.

I tried not to scream so I wouldn't scare my guys. That was harder than it sounds. Forty pounds of homemade explosives will do some damage. But I didn't want that to be their last memory of me. I just focused on breathing. If I kept breathing, they couldn't take my life.

**

It wasn't even supposed to be my day on patrol. I'd talked to the lieutenant the day before, told him flat out that we needed a break. My squad had been getting run ragged. It seemed like every time shit went down, they'd call for us specifically. But One Bravo was up in a hide position, watching a tracked mine that a farmer had been moving around from the market.

Command couldn't see the bag from their position, so they sent us to mark it. Simple enough mission. Just pretend we're on regular patrol so we wouldn't tip our hand. I only took half my squad—I learned early on that smaller groups moved better in that environment.

When we got there, everything felt wrong. The Afghan guy who ran the little snack shop wasn't there. I always allowed my guys to go in there and buy little things—chips, soda. It was my way of trying to show the local populace that we were there to help, not raid and take shit for free. I thought I'd build a rapport with the guy. In retrospect, he knew something was up, which was why he was gone that day. But I didn't know that then, and didn't have time to think about it too much. We had a job to do.

Turns out, that feeling was right. Even when you do everything right, check every step, sometimes it doesn't matter.

The blast was forty pounds of homemade explosives. These guys had started using less metal, but more plastic and wood in their IEDs. Harder to detect.

I thought I'd gotten pretty good at spotting IEDs . . . as much as that's possible, anyway. The longer you're in the AO (Area of Operation), the more you figure out what's going on around you.

I learned to study civilians, watching the paths they walked on and the spots they'd intentionally avoid. If they ducked off the path and got back on, sometimes we'd call it in (like we were supposed to) and have the experts come in and detonate them. Other times, I got impatient, and I would go and dig them up myself, although I admit that was a little sketchy at times. We'd also developed a pretty good relationship with the locals because we were there to help them, not hurt them, so sometimes the civilians would come to us at night and tell us where an IED was.

The one that finally got me, though? That one was a surprise.

What I'm about to tell you is a little bit of what I remember and a lot of what my guys told me after the fact. Turns out, when you get blown to pieces, recalling a play-by-play isn't so easy. I know that in the blast, both my arms got blown off. So did my right leg. I know I screamed a couple of times before making myself stop. *I don't want this to be the last thing my guys hear from me*, I thought. My left leg was twisted so badly that the guys told me afterward they thought it was my right. The doctors would have to cut it off later, but there in the field, the first thing was to stop the bleeding and give me a clear airway. My corpsman did the trach. Harrigan and Camarillo got the tourniquets on fast as hell, I later learned. Thomas was on the radio. I kept thinking: *just keep breathing.* I repeated it over and over in my head. If I kept breathing, they couldn't take my life. My squad's response was perfect—everything we'd trained for, they executed. At the hospital, I discovered that's what kept me alive—they got the tourniquets on tight enough that I didn't bleed out, and the tracheotomy in the field made the difference.

I remember my corpsman kept slapping my face to check if I was still with them. I managed to tell him if he hit me again, I was gonna reach up and punch him. He said they were making sure I was alive. I couldn't see much of what the squad was doing at that point. Somewhere in my head, I knew these guys would carry this moment with them, and I didn't want to make it worse.

The next six minutes felt like six years. I was in and out of consciousness, but I remember bits and pieces. The sound of my guys moving fast, doing what they were trained to do. The taste of blood and dust. The medevac blades getting closer.

Being carried across those bumpy fields to the helicopter was hell. Every bump sent pain shooting through what was left of me. But I kept thinking: *just keep breathing.*

Once they loaded me on the bird, I knew I'd make it. That's the last clear memory I have before everything went fuzzy. They told me later I was still conscious at Camp Dwyer, still answering questions while they worked on me. To this day, I can't remember any of that.

The mind has a way of protecting itself from some things.

**

The funny thing is that I know who built and placed the IED that blew me up. We'd raided his house twice before and found nothing. After the accident, his father—who was in Pakistan—came back and made his son turn himself in. You know what? I'm not mad at the guy. We were at war. It was his agenda versus our agenda; we were both just doing what we thought we were supposed to do. We made our choices, and changed lives with them. I'm more mad at the guy in the snack shop. Hell, he could've warned me.

Maybe, in his absence, he was and I just didn't see it in time.

In the immediate aftermath, they flew me from Dwyer to Bastion, then to Landstuhl, Germany. The doctors didn't think I'd make it and flew my family out to Germany after three days because they figured I was done. I was in and out on ketamine.

That stuff gives you weird dreams . . . except maybe they weren't dreams.

If there is a hell, I was there. The nurses were taping me to the wall, lighting me on fire, cutting me into cubes, tying me to a chair, and putting it over heat lamps, leaving me with my insides exposed to dry out. I believed the doctors and nurses were hurting me in those dreams, or real life . . . it was all blending together. When I finally came fully awake in the ICU at Bethesda Naval Hospital, I didn't trust any of the nurses. I told them all to get out and wouldn't let them touch me. That's what those ketamine dreams will do to you.

The first thing I remember clearly was asking for a beer. Two doctors actually went out and bought me a six-pack of Corona. I had a trach in my throat and wasn't supposed to be drinking anything, but they let me have a few sips anyway. I took one drink and said, "Nope, water." My mouth was so dry that it didn't quench my thirst at all. My mouth felt like it was full of sand, and all I wanted was water. That's when I asked if anyone else had gotten hurt. Always your first thought as a leader: *check on your men.*

I didn't know it then, but my family was going through their own kind of hell. The Marine Corps had sent someone to the house to notify them, and my little brother was the only one home when they showed up. They wouldn't tell him anything, just said they needed to speak to my mom. By the time she got there, cars were parked all over the yard—people had just jumped out and ran inside when they heard.

My mom, dad, and wife got emergency passports to fly to Germany. They didn't know if they'd make it in time. Everyone handles that kind of news differently. Some get quiet, some get angry, some try to take control. Put all that in a hospital room with someone fighting for their life—there's bound to be friction.

The squad carried their own weight from it. Those two Marines who saved me with the tourniquets . . . that quick action kept me alive, sure, but they have to live with those images, and the unit had to complete the deployment without me. They had to keep running missions, taking fire, and pushing forward. That's what Marines do.

**

The first time I woke up properly in the ICU, I couldn't move my hands but I could feel them—or thought I could. I asked my wife for her computer because I wanted to type something, and she had to tell me, "Honey, you don't have your hands." The mind's funny like that—it takes time to process reality.

They made me a Purple Heart presentation while I was still hooked up to tubes because I was so touch-and-go. The Commandant and Sergeant Major of the Marine Corps came themselves. It was fucking surreal to think those types of men would take the time out of their day to come to me. It showed me their tremendous character—either that, or they were doing it because everybody thought I was going to die. I had a choice about what to believe, and I chose the first option. Something inside me just knew: I was going to live.

Fighting infection was the real battle in those early days. Sand, dirt, debris—all of it was trying to kill me from the inside. The corpsmen were relentless with cleaning and antibiotics. One day, a young corpsman came to my rescue when my nether regions were sticking to my leg in the thick heat. He came in with a cold, wet rag to help, cradling that area so I could breathe a momentary sigh of relief. Talk about sacrifice! I was stuck, immobile with drains in my legs and skin grafts to monitor. Ultimately, the hardest part wasn't physical; it was watching my family try to cope.

My wife and mom both wanted to be the one who took care of me. Both had the best intentions, but they'd compete over who got to feed me, or who got to help with basic needs. I'd be lying there like some prize to be won, thinking "Come on, there's enough of me that needs help to go around." But when you're stuck in a hospital bed, you can't exactly mediate.

The ICU rules said only two people at a time in the room. My wife wouldn't leave—fair enough, she had medical power of attorney. But that meant my brothers and sisters had to rotate through that second spot. Everyone wanted time, everyone was worried, and everyone processed differently. World War III in a hospital room.

There I was, trying not to scream from pain, and I had to play United Nations between my family members. Classic middle child role—trying to keep peace from a hospital bed. "Mom, stop yelling at Crystal. Crystal, stop yelling at my mom." But I couldn't really focus on their drama. My mission was simple: survive.

The first chance I got to use a phone, I had them patch me through to my squad even though I still had my trach in. They were still in-country, running missions. I had no idea how I would be able to reach them, but a satellite phone worked, thank God. Being able to call them was a surprise for me, and I desperately wanted it to work out. I didn't want them worrying. I needed them focused on staying alive. When I finally got patched through, I tried to keep it light. Marines handle heavy stuff with humor—always have.

My voice was rough from the trach but I managed to tell them I was fine, and cracked a few jokes. I told them they'd better not screw up all my hard work training them. Made them promise we'd meet at homecoming. Leadership doesn't stop just because you're laid up in a hospital bed.

Looking back now, I think that call meant more to me than it did to them. Hearing their voices, knowing they were okay was what I needed. Squad leaders never stop worrying about their men. It doesn't matter if you're in the fight or flat on your back in a hospital bed. Those are your guys.

**

The doctors kept saying I was beating all odds, but it didn't feel that way from where I was lying. Each day was its own battle—fighting infection, enduring procedures, and trying to piece together what my life would look like now. The future was a massive question mark hovering over my hospital bed.

But Marines are trained to focus on the mission in front of them. My mission was simple: heal enough to get the hell out of that ICU. Get strong enough to start rehab. Learn to live in this new reality. The same drive that got me through boot camp, and pushed me to be a better squad leader—that's what I needed now. I didn't give myself time to feel depressed or angry about losing my limbs. Reality hit me lying in that hospital bed. I remember thinking: *How am I going to take care of anybody? How am I going to have a "normal" life? What the hell is any of this going to look like?* I knew I had a decision to make then: lie there and feel sorry for myself, or get up and do something about it. I broke it down into short-term tests: get through this surgery, get through that therapy session, then the next and the next and the next. I didn't allow myself to think long-term—a choice that got me through then, but a choice that would come back to haunt me later when I got home and the grief I'd never processed crept in, overtaking me.

But that was later. Back then, I just wanted out of that damn hospital. Finally, it was time to meet my squad at Camp Lejeune. The hospital didn't want to let me go, but I was not taking no for an answer. They arranged for me to fly to a private air base without letting any media know, my case manager and my nurses coming with me. The guys had no idea I was coming.

When I saw them, the emotions were overwhelming. I was still in the wheelchair then, but just being there, seeing their faces, and talking to their parents meant everything. I got to tell Harrigan's parents that he saved my life, shaker and all. He'd never told them, but that didn't surprise me—he's a humble guy. These were my boys, my brothers. We'd been through hell together, and even though I was different now, that brotherhood hadn't changed.

Looking back, those first days after the explosion were just the beginning. The easy part, if you can believe that. Because surviving the blast was mostly up to the docs and my training. But learning to live again, and figuring out who I was going to be without arms and legs—that battle was all on me.

I'd survived the explosion. But my real battle was just beginning.

Chapter 6

FIGHTING BACK

2010 - WALTER REED NATIONAL MILITARY MEDICAL CENTER

I feel like a goddamn baby deer on ice.

That's what I thought the first time they put me in prosthetic legs with knees, in a harness attached to a track in the ceiling so they could help keep me steady while I learned to balance. The prosthetists kept asking me stupid questions like "How does that feel?" and "Are you leaning forward or backward?" Hell if I knew—I'd never done this before. What was I supposed to compare it to? I was just trying to stand up without falling on my ass.

But then they made me 6'2" and I had to laugh. "Whoa fellas, the air's way too thin up here," I told them. "I was never this tall." They'd have to adjust the height, but first I had to learn to walk in these things.

**

The staff at Walter Reed had been warning me from the beginning that healing was going to be a long process. After the amputations, I had to wait two months before I could bear weight, even in "shortie legs." I had a hell of a lot to learn, everything from putting on the liners for my prosthetics to basic daily tasks that I used to do without thinking twice. The liners were this gel-like material that would stick to my skin and help keep the prosthetics attached. Getting them on was like a damn puzzle at first. Imagine trying to pull on a tight sock with no hands. I'd sit there for what felt like forever just trying to get the stupid things positioned right. It felt like putting on body armor . . . hell, my arm alone weighs seven pounds, if you can imagine.

And the daily stuff—man, that's where your patience really gets tested. The first time I tried putting on a hoodie, I got stuck in the thing. I couldn't get my head through the hole, and my arms got all tangled up. I probably looked like a baby being born. The occupational therapists would try to get me to do things like make a sandwich, and I'd joke, "That's what my wife is for!" But they weren't having any of that. They wanted me to figure it out myself.

For the more detailed tasks, they gave me specialized prosthetics, like a hook attachment that was supposed to help with fine motor control. The problem was, I kept breaking everything. For a while I was carrying around a backpack full of spare hands and tools and different attachments because stuff would break so often. But I knew I had to keep at it. This was my new normal, and there wasn't any other option but to figure it out.

During this time, I was living in an apartment close to the hospital that was not handicapped accessible. The bathroom doors were too narrow, and I had to use a transfer chair to get from the doorway to the toilet. It was around this time that I learned that the world is also not handicap accessible—or at least Washington D.C. wasn't. There are steps, curbs, and so many roadblocks everywhere. Historic buildings can't be modified, and everything is a hazard when you're walking on prosthetic legs. If your hands don't work yet, doorknobs are your enemy. These are the small things nobody pays attention to because they don't have to. I made it a point to be in my legs and out of my wheelchair as much as possible—something I still do today because it gives me a sense of independence. It makes me feel like Todd, or what's left of him.

One day at OT—I remember it was Halloween, or close to it, because there were pumpkins and candy—Frank Siller from the nonprofit Tunnel to Towers came to Walter Reed.

"We're going to build you a house, Todd," he told me. "One that's fully accessible to you."

I exploded with gratitude on the inside. They had no idea the questions that answered for me that I hadn't been able to verbalize—as a husband, as a provider, I had no idea where we were going to live after the hospital, no idea what life was going to look like. Then, suddenly, I did.

The physical recovery part had its own special brand of fun and also hell—a full-time job. It was complicated. Getting fitted for prosthetics isn't like trying on a new pair of shoes. They'd adjust the height, tweak the balance, mess with all these different settings trying to get everything just right. My left arm was trickier than my legs. I had to learn a whole system of muscle movements to control it. Flick my wrist up, it rotates one way. Flick it down, it goes the other direction. Do it slowly and the hand would open or close. Sounds simple enough until you're actually trying to do it. Half the time I'd end up with these random movements I didn't mean to make because the sensors were picking up the wrong signals. Once, I was talking to my grandma and had a beer in my hand, and all of the sudden my hand started spinning, dumping beer into her purse. I didn't even notice.

"Honey," she said. "Um, your hand . . ."

"Goddamnit," I said, laughing and feeling awful at the same time. "Sorry Grandma!"

Then there was the core strength work. Hours on the mat doing sit-ups, leg lifts with weights strapped on, anything to build back up the muscle I'd lost. They had me in physical therapy probably most of the day, then occupational therapy after that. By the end of each day I was physically and mentally drained. There are only so many Legos you can build in OT before you lose it. But I couldn't let myself slow down. I had to keep pushing.

The doctors had said this whole process would take four years, minimum. One year per limb lost. Four years before I'd be walking, functioning somewhat normally, and taking care of myself. But I had other plans. I threw myself into every bit of therapy they gave me. If they wanted me to do ten reps of something, I'd do twenty. When everyone else was sleeping, I'd be up doing extra exercises.

My occupational therapist, Joe Butkus, wouldn't let anyone help me with anything. If someone tried to step in, he'd be like "Nope, nope, don't help him. Let him figure it out." At first it was frustrating as hell, but looking back, that's exactly what I needed. He made me go through the chow line by myself one time right after I got a new prosthetic hand. I ended up crushing my cup and getting Sprite all over myself. Way to go, Joe. But that was his way of teaching me: throw me in the deep end and make me swim.

All that work paid off though. Instead of four years, I was up and moving in eighteen months. Later, I'd even run a 5K on the tenth anniversary of 9/11 for Tunnel to Towers and another one for Hopes and Possibilities. That last one, I got first place in my category: quadruple amputee. Probably because I was the only one in that category, but hey . . . you learn to take what you get. None of that would happen until I could get my ass out of the hospital, though, which proved harder than I thought. Despite all my requests, my paperwork kept getting "backed up." I later learned they were holding off on signing it because apparently I was an inspiration to the other guys at Walter Reed, and they wanted me to keep coming in. If they'd just asked me to stay and help, I probably would have . . . but they didn't, and I wanted to get on with my life. Finally, a politician came, and the paperwork miraculously started moving. I was medically retired from the Marine Corps with an honorable discharge.

While all this physical healing was going on, I wasn't really dealing with the mental side. I was so focused on the mission—learn to walk, figure out the prosthetics, get moving again—that I didn't take time to process everything that had happened. I was still in Marine mode, pushing through, and staying strong for my family. They were all smiling watching me hit these milestones, and I wanted to keep it that way.

At the beginning of my stay in the ICU and the first few weeks after, they had tried to send in a psychologist to talk to me. *Oh fuck no,* I thought. *I don't need this.* I'd kick them out of the room. Looking back, I realize I probably wasn't very kind and they were just trying to do their job, but I was doing all I knew how to do: be Marine tough.

But not grieving and not really dealing with losing my arms and legs would come back to bite me later. Everyone else was going through their own stress about the situation. I figured they didn't need mine on top of it. So I just kept pushing forward, focusing on the physical challenges, and tried to be the Marine they expected me to be.

And I felt like that Marine, especially coming home to St. Louis. Tunnel to Towers was still trying to find land to build the house, so I was staying with my sister. When we landed and I stepped off the plane in my shortie legs, the airport was packed. Family everywhere, news cameras rolling, Channel Five in my face. It was a huge homecoming ceremony. The Patriot Guard had shown up with their bikes, probably about 800 of them lining Highway 55. On the way out, I rode with my grandpa in his Corvette, leading the way from the airport.

It was a lot to take in. Part of me was proud to show everyone I was up and walking again, that I was still fighting. Part of me was happy to be in a place I was familiar with—not a desert, not a hospital. But another part of me was terrified of what was going to come next.

I hadn't moved to the Lake of the Ozarks yet, a place where even more would fall apart.

Chapter 7

THE SECOND BATTLE

2011 - LAKE OF THE OZARKS, MISSOURI

That first morning alone at the Lake of the Ozarks, I stood looking out over the water. A beautiful view. You could see for miles—pristine shoreline stretched out like something from a postcard. The house Tunnel to Towers built for me was perfect on paper, and I was so grateful to be only the second wounded veteran they'd ever built for. Everything was automated, everything was adapted. Doors opened with a push button, lights knew when I entered a room, and the kitchen set up so a guy with no hands could actually use it.

Funny how something that's supposed to be paradise can feel like a cage. Crystal was visiting family—the first time she'd left since we moved in. Those empty rooms echoed. Automatic doors whirred open and shut, lights flicked on and off by themselves, and each sound hammered home everything I couldn't do anymore.

Standing there watching boats cut through the water, I felt more trapped than I ever did in Afghanistan. At least there I had a mission, and knew what I was supposed to be doing. Here? Just me and a house full of machines trying their best to make up for what I'd lost.

Turns out you can feel pretty damn claustrophobic even with all that open water in front of you.

They tell you money can buy happiness, but I was fine when it came to money thanks to the support of a few nonprofits and generous donations. But the emptiness inside couldn't be replaced by any of that. It didn't have a price tag.

**

Moving to the lake seemed like it made sense at first. Split the difference between Crystal's family in Kansas City and mine in St. Louis. Plus, we both knew the area from growing up and had good memories there. Getting there was a labor of love: Gary Sinise and the Lt. Dan Band held a concert in St. Louis to raise money for the Tunnel to Towers build. We rented a little duplex while it was going up and had a little input on the design, but not much—they knew what they were doing. I was happy to be able to drive at that time; as soon as I was out of the hospital, I had my truck modified with hand controls, a steering knob, and a device that allows me to use my blinkers and other functions. Kind of funny, really—it would play "do re mi fa so la ti do." *Do* was the left blinker, *re* the right, *me* the horn, *fa* the windshield wipers . . . you get the picture. Inside, smart home technology meant I wouldn't have to rely on people for every little thing. That was the theory anyway.

Reality's a bitch, though.

Those first few months, we gave it our best shot. I learned all the house's systems, tried to get into a routine. Crystal handled what the automation couldn't. We had a steady stream of visitors at first . . . family, friends, people wanting to see the place. But that slowed down. Gaps between visits got longer.

You don't realize how much time weighs on you until you've got nothing but time. In the Marines, every day had a point: train, patrol, keep your guys alive. Even in rehab, I had goals: learn to walk again, figure out these prosthetics, or push through one more session. But now? Days just bled together. I couldn't even mow my own damn lawn. I had to watch other people do everything while I sat in my fancy automated house, feeling useless.

If you spend enough time being useless, you start questioning who you even are anymore. I'd been good at construction before the Marines. Then, I was good at being a Marine. Now, I couldn't make a fucking sandwich without it taking two hours, and Crystal would end up having to clean up whatever mess I made, anyway.

That house started to feel like it was mocking me. All that technology, all those adaptations . . . they weren't making me independent. They were just spotlighting everything I couldn't do. Every automatic door, every voice command . . . just reminders that this was my life now. This was all I'd ever be.

When thoughts like that start circling, you've got two options: face them or find a way to make them shut up. I went with option two. Looking back, that's when things really started going downhill. But at the time, it just felt like surviving another day. Different battlefield, same mission: keep breathing. Even if you're not sure why anymore.

The funny thing about being a Marine is that it gets into your blood and becomes who you are. Not just the training or the missions, but this bone-deep certainty about your place in the world. You're the guy others count on, the one who gets shit done no matter what. Then suddenly you're not that guy anymore, and you've got no clue who you're supposed to be instead.

I tried volunteering at a gun shop just to feel useful and get out of the house, working for free because they couldn't afford to pay me. It didn't matter. I would've done anything just to have somewhere to be, something that made sense. But even that felt hollow. Like playing at having a purpose.

The worst part was all these organizations wanting to parade me around like their success story. "Look who we helped!" I'm not talking about Tunnel to Towers, the Joshua Chamberlain Society, FOCUS Marines Foundations, Gary Sinise Foundation, or Semper Fi & America's Fund. . . . let me be *very* clear—those, I was happy and grateful to help and still am. Some of the others, though? They'd put me up front at events, use me in their marketing, and make it seem like I'd be making an appearance at some fundraiser when they knew damn well I wasn't. I get it. They needed to show donors that their money was doing something. But I wasn't some mascot. I was a Marine—or used to be, anyway. At that point, I didn't know what the hell I was. You spend enough time not knowing who you are, and that void starts filling up with some dark shit. I started feeling like a burden. Crystal would tell me I wasn't, but actions speak louder than words. Every time she had to clean up after me or redo something I'd tried to do myself, it just hammered home how useless I'd become. Those were my thoughts, not her words.

Crystal was there for me from the start. Through the hospital, through rehab—she was my cheerleader, my rock. But being someone's rock gets heavy after a while. The relationship shifted without either of us really noticing at first: wife to caregiver, partner to nurse. That changes things, whether you want it to or not.

She tried her best to keep me on track. Made rules about drinking, especially when family visited. "I'm the one with the problem, not them," I'd argue. But she was right. I couldn't stop once

I started. Not that I admitted that then . . . I just got resentful and started hiding it better.

Communication broke down. We'd be in the same house but might as well have been on different planets. She'd be in one room, and I'd be in another. Whole days would go by where we barely spoke. When we did talk, it was usually about what I needed help with or what I wasn't doing right.

We tried couples therapy, which went about as well as you'd expect. I got yelled at for being too honest, which seemed like missing the whole point of therapy to me. But when I think about it now, maybe I was using honesty as a weapon. Maybe I wanted a way to push her away before she could decide to leave on her own.

Because that's what I figured would happen eventually. Who wants to be married to a guy who can't even make his own sandwich? Who needs a caregiver more than a husband? I started thinking she'd be better off without me. A self-fulfilling prophecy, I guess.

Simple things become mountains when you've got no hands. Making food? Two-hour project minimum, and that's if nothing goes wrong. Try to clean something? Crystal would just have to do it over anyway because it wasn't up to her standards. Not her fault; she likes things done right. But every time she redid something I'd struggled to accomplish, it felt like another nail in the coffin of my independence.

I started living on microwave meals. Pizza rolls, TV dinners, anything I could manage without making a mess. It wasn't healthy, and it wasn't what I should've been eating. But it was something I could do myself, so that's what I did.

The days got pretty dark. You can only fail at basic tasks so many times before it starts eating at your soul. Every dropped dish, every fumbled attempt at normal life: they just kept piling up. Building this wall of resentment between us. Me resenting needing help, her resenting having to give it. Both of us resenting what our marriage had become.

But you can't show that stuff to the outside world. We had to keep up appearances. Tunnel to Towers had built us this amazing house, so I couldn't exactly go around bad-mouthing my situation. I played the role. Smiled for the cameras, said all the right things about being blessed and grateful. Meanwhile, inside I was dying by inches.

When you're living two different lives like that—the public face and the private reality— something's got to give eventually. For me, it was the marriage. We just drifted further and further apart until there was nothing left to save. Different rooms became different lives. Different dreams, different directions. The woman who'd been my rock became a stranger living in my house.

Looking back now, I can see we were both doing our best with an impossible situation. But at the time, all I could see was everything I'd lost. The man I used to be was slipping away day by day, replaced by this angry, bitter shadow I didn't even recognize anymore. Drinking wasn't new to me. Hell, it was part of Marine culture—work hard, play hard. But this was different. This wasn't about having fun or blowing off steam. This was about making the noise in my head shut up for a while. I started drinking as soon as I woke up some days. What else was there to do? I couldn't drive yet because I didn't have an adapted vehicle at that point (that came later). I couldn't really cook or clean or do any of the normal stuff that fills a day. So I'd drink and take whatever pills I could get my hands on. Anything to numb out that voice in my head keeping score of everything I couldn't do anymore.

The fucked up part was that the more I drank, the angrier I got. All that bitterness I was trying to drown would come bubbling up instead. The drunker I got, the more those suppressed emotions would surface. I started lashing out at people and becoming someone I didn't even recognize, let alone like. Self-medicating works just enough to keep you doing it, even while it's destroying everything around you. It makes the pain bearable in the moment, but the moment keeps getting shorter and shorter, and you need more and more just to get through a day. Before you know it, you're not living anymore. Just existing from one drink, one pill, one numbing agent to the next.

It didn't help that after all that time, I wound up back living with my mom again after Crystal moved out—but this time, Mom was in *my* house. It made our relationship go downhill fast, with her being overbearing and me being a shitty person to live with. It felt like I was in my rebellious years all over again—except this time, I couldn't do a backflip.

Six years of that shit will break anyone. By 2016, I was done. Not just with the drinking or the hurt from an ended relationship, but also with the endless cycle of pretending I was okay. I was done with everything.

There's no saving me, I thought one afternoon, *I'm done. This is over.* I didn't know I was feeling that way because I didn't have a chance to grieve the loss of Nick or the loss of my limbs or the loss of life as I'd known it . . . I didn't figure any of that out until I went to FOCUS later on. I knew I probably had some PTSD, but I didn't understand myself enough yet to articulate that. At that point, I was just fucking tired of being alive.

I called the veterans' crisis hotline thinking maybe they'd have some answers. They didn't—in fact, I asked to speak to a supervisor, and nobody ever called me back. Six hours later, after it was dark and I'd been drinking a bit more, I called back and chewed them out for not returning my call when I was in crisis. After all, what good is a fucking crisis hotline if they don't help you when you're in crisis? You won't believe this, but the woman on the other end of the line told me I was too irate and that I should call back in 30 minutes. I hung up the phone with my answer: *If a veteran's suicide hotline can't help me, I'm fucked. Nobody will be able to help me.*

I made my decision right there, and I wasn't even particularly emotional about it. I just felt . . . done. Like I'd given it my best shot and failed. I figured everyone would be better off without having to take care of me anymore. My family could stop worrying. It made sense at the time.[1]

I left a note on my phone but I don't even remember what it said, in my haze. Got my gun. I was thinking ahead: I wanted my family to have one last open casket of me, so I decided not to shoot myself in the head and instead go for the chest. The bullet nicked an artery in my heart. It should've been fatal. God apparently had other plans.

My mom found me lying on the deck, and I was rushed to the hospital. Again.

When I woke up, the look in my family's eyes shattered me more than the bullet—the combination of disappointment and shock that I could have even *wanted* to take my life is what really keeps me from doing it today. After I tried to kill myself, my Marine brothers found out somehow and forced their way into my hospital room—literally. My family tried to stop them, telling them I said I didn't want visitors. They weren't having it. "We're coming in as a fire team," they said, pushing their way through the doors.

The look in their eyes when they saw me damn near broke me. "What are you doing, man?" they demanded. "All that hard work we put in to save your life and you're trying to take it?" It wasn't said out loud, but it was their faces. The guys who'd fought to save me in Afghanistan, who'd stayed by my side through recovery . . . I'd let them down.

[1] If you are a veteran who is struggling, dial 988.

I never felt smaller than I did in that moment. Here I was thinking I was just ending my own pain, and I hadn't really processed how it would impact everyone else. They'd risked their lives to save mine twice now—once in Afghanistan, once in that field hospital—and this was how I repaid them.

Lying in that hospital bed, watching what my choices had done to the people who cared about me, that's when it finally hit home. The same Marine mindset that had kept me alive in combat was killing me as a civilian. That stubborn refusal to show weakness, to admit I needed help—it was a form of pride. The Your Choices Change Lives tattoo rang my bell again, slapped me in the face. Only this time, it slapped me back awake.

I was in the hospital again. Different circumstances, same antiseptic smell. This time they had to put in a heart defibrillator. That's what a bullet through your chest will do to you. Funny how you can survive an IED that should've killed you, then nearly take yourself out six years later with your own gun.

My family's reaction was complicated: pain, anger, and fear all wrapped up in this crushing silence. Nobody really knew what to say. What do you say to someone who just tried to check out? Especially when that someone is supposed to be the tough Marine who survived impossible odds.

But my Marine brothers knew exactly what to say. No tiptoeing around, no careful words, just straight truth. They weren't about to let me hide behind that tough guy mask anymore. They made me face what I'd done, not just to myself but to everyone who cared about me.

It was different than the hospital after Afghanistan. Back then, everybody was focused on what I'd lost: my arms, my legs. This time it was about what I still had but wasn't using: people who gave a damn whether I lived or died. People who would've helped if I'd just been willing to ask.

Sometimes you've got to hit bottom before you can see clearly. Lying in that hospital bed, watching what my choices had done to the people around me—it was like this fog finally lifted. All that time I'd spent feeling useless, feeling like a burden . . . I'd been focused on everything I *couldn't* do anymore. Meanwhile, these people saw something in me worth fighting for, worth saving. Even when I couldn't see it myself.

I started thinking about that Marine mindset that had gotten me through so much. The one that said asking for help was a weakness, that I had to handle everything myself. Maybe that worked in combat, but this wasn't combat anymore. This was life. And life isn't a solo mission.

**

Looking at my reflection in the mirror—new scars on my chest joining all the old ones—it hit me: I'd survived again. Third time now if you count the IED and all the surgeries. But survival wasn't enough anymore. I had to figure out how to actually live.

The guy in the mirror wasn't the same Marine who'd led men in Afghanistan. He wasn't the same guy who'd pushed through rehab in record time, either. This was Todd 2.0, and I still didn't know exactly who that was supposed to be. But for the first time, I was starting to understand that figuring it out wasn't a one-man job.

Looking at those scars in the mirror, the old ones from fighting enemies, the new ones from fighting myself, I knew what I had to do. The hardest battle wasn't going to be learning to live without arms and legs. It was going to be learning to live without that armor I'd built up. Learning to ask for help.

Choices change lives—my mantra. My choices in Afghanistan had saved my men. My choice to fight through recovery had inspired others. Now I had another choice to make. I could keep trying to be the guy who never needed help and never showed weakness, or I could try something that scared me more than any firefight ever had: admitting I couldn't do this alone.

There in that hospital bathroom, I finally understood: sometimes the bravest choice is admitting you need help to make it through another day above the dirt.

Chapter 8

LEARNING TO ASK

2019 - AGUSTA, MISSOURI

Getting out of my buddy Jake's truck at FOCUS Marine Foundation felt like walking to my own execution. It had been eight years of telling them to fuck off, that I didn't need their help or their stupid program. Now here I was, backed into a corner by a fiancée and a best friend—and they weren't taking no for an answer.

Rock bottom has a way of changing your mind about things. But still, I had Jake's number on speed dial. I told him that if I called him, he was to come get me immediately.

The room was filled with veterans, some missing pieces like me, others carrying wounds you couldn't see. I did what I always did—found the back wall, scoped the exits, tried to look like I gave a shit about being there. Old habits don't die easy.

That first day, I kept thinking about bailing. These guys didn't know me. Didn't know what I'd been through. What the hell could they teach me that eight years of figuring shit out on my own hadn't?

But there was something different about this place. Something that cut through all my practiced bullshit.

**

I met Michelle online in December of 2016—she seemed fun and outgoing, and we vibed really well. We talked on the phone for a month before we ever met in person—at this little dive, Scotty's, which we both enjoyed—and we just clicked. From the moment I laid eyes on her, I just knew. This was it.

I'll spare you the mushy stuff, but I was in love. It was like getting a whole new family: she had two kids, ten and two at the time, and we would eventually go on to have our son Elijah. It fulfilled another dream of mine: I always wanted to be in the military (check), and I always wanted to have a family and break the cycle of the Nicely name (check).

Michelle was always pushing me to get better, to get help. And hell, I tried. I'd been in alcohol rehab facilitation through the VA, but the problem is that they put you in a mental ward for lockdown and detox before you can join the program. Underneath those neon bulbs, it all felt too structured, too sterile, too medical, too "for everybody" and not "for me." I tried an outside therapist, too, but that wasn't working, either. I always knew where she was going before she got there, asking me "How did that make you feel" over and over. No solutions, just more questions, more talking about feelings. Fuck that. No thanks, not for me.

FOCUS Marine Foundation had been trying to get me to attend their program for eight years—it's a place meant for anyone who signed on the dotted line, a place meant to re-civilize you and help you adapt to life outside of the military. It combines all the trauma stuff with real-world advice and application. I'd known about them but, surprise, surprise, had avoided them at all costs. At events, they'd leave business cards with me and the whole nine, but I was hesitant: I didn't feel like being the face of another organization. And to be honest, I was falling back into my old ways—the alcohol, the numbing.

I remember the night before FOCUS clearly: it was my birthday. I was out with a few of the guys from the Joshua Chamberlain Society—a foundation that started in St. Louis and adopts one hero at a time *for life,* quality over quantity. Those guys are like family to this day. Anything we need—a roof caves in, a gold star kid needs money for college, you name it—I'd do anything for them, and they would for me, too. That night, some guys from the foundation—John Mabry, the ex-Cardinal player, Matt Cutler, and Charlie Hildebrand—took me out and got me saucy.

"This is what we're going to do," they said. "You're going to go to FOCUS in the morning."

"Nope," I said. "I'm not doing that kumbaya shit. Plus, I'll be hungover."

"Then you can go in the afternoon," they said, clearly seeing something I didn't: that I was crumbling fast. Michelle was in on it, and had been pushing me to get help. She wasn't about to watch as I self-destructed for a second time.

I caved, ultimately. And it was for the best. Some people who were smarter than me knew I needed to be there.

When I got out of my buddy Jake's truck that next day, I wasn't feeling it. And I was honest about that. It was always okay to be honest with Jake, my best friend since fourth grade. He had been there through it all, seen me at my lowest of lows, and showed up when everybody else drifted away. If anybody was the right person to drop me off there, it was Jake. I put him on speed dial in case I needed to get away quickly, and I went in with my standard armor on: Marines don't cry. Marines don't need therapy. Marines figure it out or die trying. I'd survived getting blown up, then learned to walk in 18 months when they said it would take four years. I didn't need anyone's help.

The counselors would just look at me and say I was too smart for my own good. They'd seen right through my act: the tough Marine routine, the "I'm fine" bullshit. I'd perfected that mask over eight years. But these guys weren't buying what I was selling.

"You know what we're trying to do," they'd say when I'd try to outthink the program. Truth was, I did know. I just wasn't ready to admit I needed it. It took me watching other Marines—hard men, warriors who'd seen worse shit than me—stand up and say they needed help. That cracked something in me. These weren't weak men looking for pity. These were my brothers, speaking a language I understood. They'd been to hell and back just like me. But somewhere along the way, they'd figured out something I hadn't: sometimes the strongest thing a Marine can do is admit he can't do it alone.

Watching them, I started thinking maybe I had it backward all those years. Maybe being too hard-headed to ask for help wasn't strength at all. Maybe I was just being a stubborn asshole, letting pride get in the way of actually getting better.

**

FOCUS hit me with a question I wasn't ready for: "You know what you're willing to die for, but do you know what you're willing to live for?"

That shit knocked me on my ass. That's what kept me in the seat that day, what made me park it and listen. I'd spent years numbing myself with booze and pills, going through the motions. But an actual reason to live? I couldn't tell you. It took three or four days of that question eating at me before I figured it out: I needed to live for myself. Not for the medals, not for other people's expectations, just for me.

For so long, I'd felt like I was living a double life. I didn't like being put on some pedestal on the outside, having to only show my happy side to the world while I was drowning inside.

The real kicker was realizing I couldn't do that alone. I'd watch guys helping me with my plate at meals, and it pissed me off at first. I thought, *I've got this, leave me the fuck alone.* But they kept at it, not making a big deal, just being there. It started making me think maybe I'd been looking at help all wrong. These guys weren't treating me like I was broken. They were just Marines taking care of their own.

One day, John O'Leary showed up to speak. He's this guy who looks pretty put together from the outside until you learn he was burned on over 90 percent of his body. He began talking about losing his fingers. Everyone in the room got really quiet, giving him that sympathy look. I just started laughing. People turned and stared at me like I was the biggest asshole in the world until I held up my prosthetics and said, "Hey, I don't have hands either."

We clicked from then on. There's something about finding another guy who can laugh at the shit life has thrown at us. Every time he'd come to speak after that, he'd mention something about having a trach. I'd raise my hand and he'd go, "Okay Todd, yeah. We know." It made me realize maybe I wasn't the only one who used humor to deal with the heavy stuff.

O'Leary got through to me in a way nobody else had. I was drawn to him. Here was a guy who'd been through his own version of hell and come out the other side. Not by being the toughest or most stubborn, but by learning how to let other people in. It made me start thinking maybe I could do the same.

It wasn't like some magical moment where everything changed, but between the program showing me a different kind of strength and O'Leary showing me you could go through serious shit and still find purpose, something started shifting.

I started seeing that maybe being invincible wasn't the point. Maybe being human was enough. Seeing O'Leary use his story to help people, I started thinking, *maybe I can do that.*

The Marine Corps taught me strength meant never showing weakness. They taught me to be hard, and push through anything. Keep the Marine mindset, handle your own shit. But that thinking nearly killed me.

At FOCUS Marines Foundation, they kept telling me to "keep the Marine in your pocket." Pull him out when you need him, but don't let him run your whole life. That sounded like bullshit at first. The Marine was who I was—what else was left? But slowly I started getting it. That armor I'd built up, all that "I don't need anybody" crap, was actually making me weaker.

Real strength? That came from being able to say "I need help" without feeling like less of a man. From admitting I didn't have all the answers. It was the hardest thing I ever had to learn—harder than walking again, harder than figuring out how to wipe my own ass. But it saved my life.

**

My wife Michelle—well, she was my fiancée then—saw me differently from the start. To her, I wasn't some wounded warrior who needed fixing. Not as a quadruple amputee. She just saw me as Todd. She didn't even know I had the lake house at first. Didn't Google me or any of that shit. Just took me as I was.

But two years after I put the gun to my chest and pulled the trigger, I started slipping and going back to the bottle, back to my old ways. I still didn't know who the hell I was anymore, didn't know *why* I felt the way I did. I'd never taken the time to really sit with the fact that my arms and legs were blown off, that Nick was gone—they were facts, but I didn't let myself feel any kind of way about them. Jake was one of my only friends who kept coming around no matter what place I was in, who knew I wasn't in my right mind but stayed by my side anyway. I love that guy—he is my brother. And Michelle wasn't having it, either . . . wasn't about to watch the man she loved destroy himself. Can you blame her?

FOCUS taught me how to actually talk to her—not just the surface bullshit, but the real stuff. The dark thoughts, the frustrations, the fears . . . all the things Marines aren't supposed to admit they feel. Learning to do that shit wasn't easy. In fact, it still isn't. But it beats the hell out of pretending I'm fine when I'm not.

I had to learn it was okay to lean on Michelle sometimes. That being vulnerable with someone you trust isn't weakness; it's how you build something real. She saw me fall a thousand times and still chose to stay. She made me realize maybe I was worth more than I thought. That ended up being in my wedding vows to her: that she saw me at my worst and deserved me at my best. It was true then and it's still true now.

But it's not like this transformation happened overnight. People want to act like getting your head straight is some kind of magic fix: go to therapy, take your meds, and boom, you're cured. It doesn't work like that. In fact, it took me forever to accept that I needed medication. Marines don't take pills for feelings, right? We suck it up. But that's the thing about depression . . . you can't just muscle through it. Trust me, I tried. All that got me was a bullet in my chest and a defibrillator to show for it.

Here's the real bitch of it: the meds actually work, but your brain plays tricks on you. You start feeling good and think, "I don't need these anymore." The next thing you know, you're back in that dark place. I had to learn that one the hard way.

Now I stay on them because I know better. I know they don't make me weak . . . they make me smart enough to use the tools that keep me alive.

Another change was that after FOCUS, I had to step back from my family for a while. Not because I didn't love them, but because they kept seeing me as who I used to be. They kept waiting for old Todd to show up—the fuck-up, the drunk, the guy who couldn't get his shit together. I get it. After all, I'd put them through hell. I was changing, and I needed space to do it. They didn't get it at first and thought I was pushing them away again. I had to learn to set boundaries . . . something Marines aren't exactly known for.

The funny thing is, once I started taking care of myself, they started seeing the change. It wasn't overnight, but suddenly one person would notice, then another. Like dominos falling. Now we're closer than ever, but it's different. Healthier. They don't walk on eggshells around me anymore, and I don't feel like I have to prove anything to them.

We never had some big dramatic moment where everybody apologized and hugged it out. That's movie bullshit. Instead, we just gradually started treating each other better and learned to let the past be the past. Everyone knew what had happened; we didn't need to keep talking about it. We just needed to move forward.

It got to a point where I could be around them without needing a drink, where I could actually be present instead of just physically there. That's when I knew things were really changing: when I could sit in a room with my family and just be Todd.

Not Marine Todd.

Not quadruple amputee Todd.

Just Todd.

Just me.

Of course, I had to relearn everything after getting blown up—how to walk, how to feed myself, all that basic stuff. But learning to actually live again? That was different. I had to figure out who I was without all the armor I'd built up.

Slowly, I began accepting my limitations. *Yeah, I can't vacuum worth a damn, and Michelle's probably gonna redo it anyway. That's okay.* I found other ways to contribute, and be useful. I built systems that work for me instead of trying to do everything the "normal" way. I got a zero-turn mower, and now I take care of six acres. My clothing is altered so I can put it on without help. My prosthetics are tools, and I've got to look at them that way. Sometimes I can do things easier without them, and sometimes I need them. I will do whatever I need to do to solve the problem at hand, no pun intended. When I'm putting on deodorant, for example, I have to grab it with my left prosthetic hand and then take that off to get underneath my right arm. There's a system to getting ready that I've had to perfect over the last fifteen years, and it's still evolving, just like me.

If I run into an obstacle doing some day-to-day task that I used to take for granted—which, believe me, happens a lot more than I'd like to admit—I at least make the choice to *try*. Again, it's a choice.

Using humor helped. It still does. People see the prosthetics and don't know how to act, so I break the ice . . . like that time my hand fell off and rolled under some lady's airplane seat. When I had to ask her to grab it, I said, "Bet you've never heard that request before." She cracked up. It makes it easier for everybody when you can laugh about tough stuff.

Celebrating progress looks different now. I used to measure everything against what I could do before, but that's a losing game. Now I look at where I am versus where I was. Some days that's just getting out of bed and choosing to keep going. Other days, it's figuring out a new way to do something I thought I couldn't. Progress is progress.

**

Eight minutes. That's how long they give you to tell your story to the new FOCUS class when you're a mentor and not in the seat. Standing up there that first time, looking at those faces, I saw myself in every one of them: the anger, the pride, the pain they're trying like hell to hide.

I told them straight up about the dark times. I didn't sugarcoat the suicide attempt or the years I wasted trying to be invincible. But I also told them about coming back from that edge. About learning that the strongest thing a Marine can do is admit when he needs help.

I pay close attention to their faces when I get up there. Some of them have that same look I had, the one that screams, *This is some bullshit kumbaya crap.* But over time, you see it start to crack. You see that moment when they realize maybe, just maybe, there's another way.

It is truly a full circle moment when I am on that stage, and I've created a bond with a lot of the mentors there for the time I've been doing it—four weeks a year for the past six years, and I don't plan on stopping. But back in March of 2019, I was sitting in that chair thinking I didn't need to be there and had all the answers. Now I'm the one telling them it's okay to not be okay. That asking for help isn't weakness—it's how you survive. How you learn to live again. I tell them that this isn't some bologna bullshit—it's true, and it works. But nobody is going to do it for you. You have to make the choice.

In the program, we talk about who you are now, who you want to be, and where you want to go. Some of the vets come up afterward and want to know how I got where I am. I tell them the truth—that I didn't do it alone, and none of us do. We're

Marines, and we take care of our own. Sometimes that means letting our own take care of us. And I'll be honest with you: I never wanted to be the one people pointed to and said, "Look at this guy!" I still struggle with that. But at least now I have the tools and awareness to stop it before it goes too far.

It hit me hard the night after I spoke to the class for the first time: I'd spent eight years running from help, and there I was learning how to give it. Life's got a weird sense of humor sometimes.

The Marine Corps taught me to run toward gunfire to protect my men. But they never trained us how to fight the battles that come after, the ones inside your head. Now when I see that look in another veteran's eyes, that same pride and pain I carried for so long, I know what to do. Not just because FOCUS taught me, but because I lived it.

Some guys ask me if I regret taking so long to get help. And sure, I could've avoided a lot of shit if I'd done it sooner. But I tell them the truth: those eight years, as fucked up as they were, are part of my story now. They're part of what lets me connect with guys who think they're too tough, too stubborn, or too far gone to be helped.

FOCUS Marine Foundation showed me that asking for help isn't the end of being a Marine—it's just the beginning of being a different kind of strong.

Epilogue

STILL CHOOSING

2025 - ST. LOUIS, MISSOURI

The sun was barely up, and I heard my son stirring. That kid's got more energy than a platoon of Marines. My wife was already smiling—she knew I was about to get ambushed by forty pounds of pure chaos.

I caught my reflection as I put on my prosthetics. There's a whole order of operations that took years to figure out. But it's not the physical routine that matters. It's the mental one. Some days it still hits hard, seeing what's left of me. But then I think about everything that's happened since that day in Afghanistan. All the lives that have crossed mine. All the brothers I've helped find their way back.

Above the dirt another day. That's not just luck anymore—that's a responsibility. Along the way, I've learned that we can't be hard on ourselves when we make a mistake. I have been my own worst enemy—at my lowest, no one could punish me more than I was punishing myself.

You know, sometimes the world has a funny way of telling you which person you're supposed to be. I thought I knew everything at 26 in the military. I was having the time of my life, stepping into leadership before stepping on that damn IED. Then, everything changed.

I've got a family who needs me now, vets who need to hear there's hope and a mission that's bigger than my own struggles. I didn't ask for any of these roles, nor do I have to fill them. But I make a decision to get it done.

Times change. People change. And so can your choices.

Choices change lives.

**

I've learned to be grateful for the simple stuff—being able to play with my kid, make my wife laugh, and help another vet find his way back. Things I used to take for granted, or didn't think mattered.

I still use humor like a weapon. I have to. Life's too heavy if you can't laugh at it sometimes. And family comes first now. It took losing almost everything to figure that out. My son doesn't care that I can't throw a baseball like other dads. He just wants me present, engaged, and actually there instead of numbing myself with a bottle. That's a choice I make every day: to be the father my kids deserve, not the one I'm afraid of being.

Morning routines, prosthetics, family time—they're all choices now. Every single one. Good days, bad days, they all start with deciding who I want to be today. Marine stubbornness finally found a good use: choosing to keep pushing forward, no matter what.

And look, I'm not perfect. I have bad days . . . *really* bad days, if you want the truth. Some mornings I wake up and the demons are there waiting, snarling at me. There's no other way to say it: it's fucking hell. If you're going through something like that, let me tell you, I get it. I know what that feels like. But I also know one thing I've learned over all these years with these scars, the ones you can see and the ones you can't: leave the past in the past. If you have a bad day, make the choice to get up again tomorrow and keep trying. It's easy to say and hard to do, but at the end of the day, it is the choice we need to make. And trust me, it's worth it.

Things really opened up for me when I learned it was okay to ask for help, to be able to talk about what happened. That choice changed my life. I was able to bring my family back, and mend bridges I'd burned. **If you're reading this and think you might need help, you're right. Get it**. It doesn't have to be the organizations that worked for me. There's always at least one person for everybody—at least one person you can talk to, at least one person who *needs* you to stick around.

I never thought I'd be a mentor. Hell, for years I was the guy who needed mentoring but was too stubborn to admit it. Now I'm at FOCUS Marines Foundation every session, watching guys come in with that same look I had—like they'd rather be back in combat than dealing with their shit.

On Tuesdays, I use my eight minutes in front of the group to show them there's another way. And Honor Guard hits differently now. I used to think it was just about the ceremony, the tradition. But it's more than that. Some of these guys we lay to rest, they've got one car in the lot. One person to remember them. But they wrote a blank check to this country, the same as any of us. The least I can do is make sure they're sent off right.

I started doing some public speaking, and even got a podcast going called When Does The FKRY End. My first episodes were just me and my brother-in-law getting drunk, talking shit—lots of veteran humor. Now we're interviewing older vets and getting their stories down before they're gone. We even had a guy on who did damage assessment at Hiroshima. That's the kind of history that needs saving. I believe in the value of archiving more veteran stories, and not the sanitized history book version . . . the real deal. The fear, the brotherhood, the stuff that keeps you up at night. The stuff that makes us human. That's what the next generation needs to hear.

Lately, I've been thinking a lot about what I want to leave behind not just for my kids, but for all the vets coming up behind us. We're losing World War II guys every day. Korea and Vietnam vets aren't far behind. Their stories matter, not just the battles, but what it was really like on the ground.

My goal is to expand what we're doing at FOCUS Marine Foundation and offer more outreach, more support programs. There are too many guys still thinking they have to fight their battles alone and too many families not knowing how to help. That shit stops with us.

At the end of the day, my family is my core focus. My son and my two stepkids are growing up and watching how I handle my struggles. I'm proud to teach them that it's okay to ask for help, to not have all the answers. That's a legacy worth building.

Everything I do now comes back to one thing: choices change lives. Every vet I help, every story we save, every funeral we honor . . . it all matters. All part of something bigger than me.

Tomorrow's not promised—I know that pretty damn well. But today and every day, I get to choose.

I'm still choosing to live.

Bootcamp Graduation

Preparing for patrol in the Helmland Province

Active Duty

Two not pictured: LCPL Sousa; LCPL Gizienski

Back row L to R: LCPL Listle; LCPL Thomas; CPL Marsden; LCPL Camarillo; LCPL Barnes; Corpman "Doc" Pickens; LCPL Turvey; LCPL Ploog

Front row: Cpl. Nicely; LCPL Geddis; LCPL Ortiz; LCPL Harrigan

LCPL Nick Hand pretending to be our platoon sergeant

One of the last photos taken of LCPL Nick Hand in Afghanistan

Homecoming with my squad. As I promised,
I would be there.

Purple Heart presentation in my hospital bed

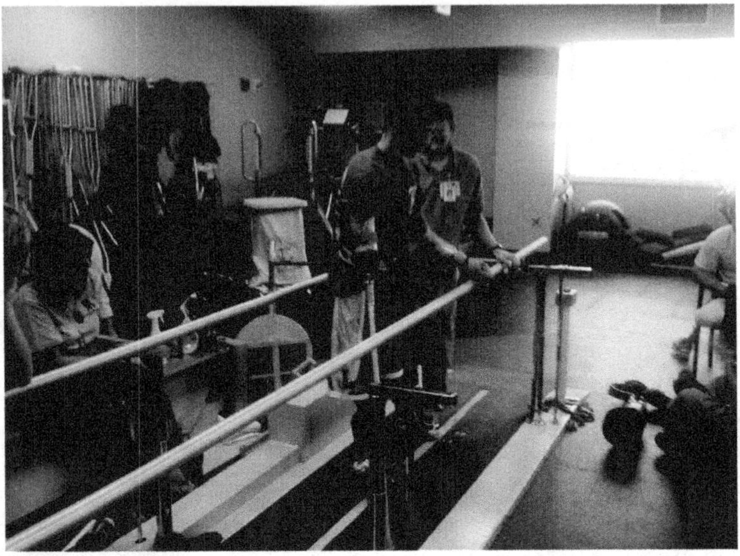

A day at physical therapy, learning to walk

Picture taken by Devin Mitchell of the Veterans Vision Project

NICHOLAS J
HAND
LCPL
US MARINE CORPS
AFGHANISTAN
MAY 20 1989
NOV 22 2009
PH KIA
SON & BROTHER

My son Elijah and I, visiting my hero

My family with Elijah being held by my wife, Michelle, with Bella in the middle and Evan on the right

Marine Corps League 707 Honor Guard crew for funerals at Jefferson Barracks National Cemetery

Another man who inspires me, by the name of John O'Leary

Speaking at FOCUS Marines

Summary of Action

Corporal Todd A. Nicely was in receipt of imminent danger pay during this period.

Corporal Todd A. Nicely is enthusiastically recommended for the Navy and Marine Corps Achievement Medal with Combat Distinguishing Device for his actions while serving as Squad Leader, 1st Squad, 1st Platoon, Company F, 2d Battalion, 2d Marine Regiment, Regimental Combat Team-7, Marine Expeditionary Brigade-Afghanistan in support of combat operations in Garmsir District, Helmand Province, Afghanistan from 11 November 2009 to 10 January 2010.

On 12 November 2009, Corporal Nicelys squad exited friendly lines at approximately 0530D hours to set in an over-watch position for 3rd Squads patrol later that day. At 0615D hours, 1st Squad reached their over-watch position and set in to observe any activity towards the South and West. At 1030D hours, the 3rd Squad moved out of the patrol base and began to cross Route Dolphins to move towards the West. At approximately the same time as the 3rd Squads movement, Corporal Nicely noticed that civilians were fleeing to the East. The 3rd Squad began taking enemy small arms fire from the South and Southwest while attempting to cross Route Dolphins and was pinned down by the enemy in a building on the eastern side of Route Dolphins. The 1st Squad quickly gained positive identification and began to suppress the enemy fighters in two different buildings that were located approximately 200 meters to their South. Corporal Nicely quickly decided to move his squad South to close on the enemys fighting position. The only way to reach the buildings from which they were receiving fire from was to rush through an open field. Corporal Nicely immediately started to rush through the open terrain, and the rest of his squad followed.

Upon reaching the first building from which they were receiving contact from, the 1st squad quickly swept the area and searched the compound, finding nothing significant. Corporal Nicely assumed the enemy fighters had moved further South to occupy another fighting position. While completing the search of the building, Corporal Nicely received information from Patrol Base Lakaris Combat Operations Center that there was suspicious activity near a mosque just Southwest of their position. Corporal Nicely decided to move his squad further South in order to investigate the information that he had received and continued to close on the enemy. The 1st squad continued to move South by bounding towards three buildings and set into a security posture to the South and West. As they reached the compounds, the squad again sustained enemy fire from the South and West. Corporal Nicely quickly gained positive identification of the enemy firing position and gave firing commands to his machinegun team and fire team leaders. The squad quickly gained fire superiority and the enemy fire ceased. At the same time, an adjacent platoon was maneuvering a squad to the East of 1st squads position. Corporal Nicely communicated with the squad leader from the adjacent squad and directed him on where the enemy fighters were firing from. Because of Corporal Nicelys understanding of the concept of mutual support, he and the other squad leader quickly closed on the enemy position and forced the enemy fighters to withdraw to the South. During the search of the area, Corporal Nicelys squad found multiple ammunition caches that were left by the enemy.

On 23 November 2009, Corporal Nicelys squad was assigned as a quick reaction force for a squad from 4th platoon that was patrolling in Mohbi Khan. Before his squad could depart friendly lines, they began receiving enemy small arms fire. Corporal Nicely led his squad out of the patrol base while receiving fire in order to assist the squad from 4th platoon. Corporal Nicely maneuvered his squad to the northwest towards a tree line in order to observe the activity in the West. Because 4th platoon was calling in indirect fire support, Corporal Nicelys squad had to move South to stay out of the gun target line. During his squads movement, Corporal Nicely continued to communicate with the squad leader from 4th platoon in order to support them as best as possible. A squad leader from 4th platoon said that they were receiving effective fire from a building that was located to the west of Corporal Nicelys position. Corporal Nicely had to slow the speed of his squads movement because artillery rounds were impacting within 200 of their position. Once the indirect fire ceased, Corporal Nicely began to bound by teams to the West and headed towards the building that the 4th platoon was receiving fire from. As Corporal Nicelys squad closed on the building, they began receiving heavy enemy fire. Because of Corporal Nicelys maneuver tactics, his machinegun team was in position to rapidly suppress the enemy fire. The suppressive fire allowed the rest of his squad to move forward towards the enemy position. As Corporal Nicely reached the building where the enemy had been firing from, he quickly conducted a search of the area but the enemy had already withdrawn. After multiple volleys of mortar fire, Corporal Nicely then coordinated with his adjacent squad leader to conduct a detailed clear of the tree lines where the enemy had established new firing positions. While no enemy fighters were located, enemy radio traffic, local national reporting, and Afghan National Directorate of Securitys intelligence assessments confirmed that Corporal Nicelys squad and his adjacent squad killed six enemy fighters in this engagement.

While on hundreds of subsequent patrols in one of the most dangerous areas in southern Afghanistan, Corporal Nicely and his squad found five improvised explosive devices and three ammunition caches. On a patrol on 23 December 2009, Corporal Nicely searched a building that was a possible staging area for Taliban fighters. In the compound, Corporal Nicely found important documents with over fifty names associated with the Taliban and phone numbers that they had used. His discovery led to the constant pursuit of high value insurgents.

Corporal Nicelys relentless pursuit of the enemy during this period has had a profound impact on enabling stability to take hold in what was once thought of as the most dangerous area in southern Helmand Province. Corporal Nicelys leadership and determination to continuously pressure, punish, and to pursue the enemy are deserving of a Navy and Marine Corps Achievement Medal with Combat Distinguishing Device.

Acknowledgments

Thank you to my mother for always being there for me even though I know I've been an ass at times. I love you.

Thank you to the organizations that have changed my life in more ways than one, including: Joshua Chamberlain Society, Tunnel to Towers, Gary Sinise Foundation, FOCUS Marines Foundation, segs4vets, and Semper Fi & America's Fund.

Thank you to my fellow Marines who did what they needed to do that day in 2010 and to my medical team, including Joe Butkus (OT) and Greg Loomis (PT), who did what they had to do for so many years after. I know I wasn't always the easiest patient. I owe you guys.

Thank you to my wife, Michelle, and my best friend, Jake, for being there on good days and bad days and being smarter than I'll ever be by pushing me to go to FOCUS. Thanks, too, to Kim Bradley from Semper Fi & America's Fund for nudging me toward FOCUS with patience and persistence. You all saved me.

Thank you to my mentor and friend, John O'Leary, for giving me the push I needed to make this book happen and for being an all-around hell of a friend.

Thank you to my editor, Jessica Burdg, and the rest of my editorial and publishing team for helping me share my story.

Thank you to those who have been there for me emotionally over the years. It would take me all day to name everyone, but I'm lucky the list is so long. You know who you are.

About The Author

Todd Nicely is a retired United States Marine Corps Corporal and one of only five surviving quadruple amputees from the wars in Iraq and Afghanistan. Raised in Missouri, Nicely enlisted in the Marines and served with distinction in Afghanistan, where he sustained life-altering injuries from an improvised explosive device.

His journey through trauma, addiction, and recovery has made him a sought-after advocate for wounded veterans and mental health awareness. Nicely's story has been featured in national media, and he regularly speaks to audiences about resilience, the power of choice, and overcoming adversity. Your Choices Change Lives is his unflinching memoir, offering hope and inspiration to anyone facing hardship.

Made in the USA
Monee, IL
06 March 2026

44947937R10066